Naturalists' Handbooks 9

Animals on Seaweed

PETER J. HAYWARD
Research Fellow, Marine Research Group, University College of Swansea

*With colour plates by
Arthur Byrne
and illustrations by
the author*

Richmond Publishing Co. Ltd.

*Orchard Road
Richmond Surrey TW9 4PD England*

Series editors
S. A. Corbet and R. H. L. Disney
Advisory board:
J. W. L. Beament, V. K. Brown,
J. A. Hammond, A. E. Stubbs

Published by The Richmond Publishing Co. Ltd.,
Orchard Road, Richmond, Surrey TW9 4PD
Telephone 01-878 8955

© The Richmond Publishing Co. Ltd. 1988
© Line illustrations and text Peter J. Hayward 1988
© Colour illustrations Arthur Byrne 1988

ISBN 0 85546 265 5 Paper
ISBN 0 85546 266 3 Hardcovers

Printed in Great Britain by Cambrian News,
Aberystwyth, Wales

Editors' preface

Students at school or university, and others without a university training in biology, may have the opportunity and inclination to study local natural history but lack the knowledge to do so in a confident and productive way. The books in this series offer them the information and ideas needed to plan an investigation, and the practical guidance to carry it out. They draw attention to regions on the frontiers of current knowledge where amateur studies have much to offer. We hope readers will derive as much satisfaction from their biological explorations as we have done.

The keys are an important feature of the books. Even in Britain, the identification of many groups remains a barrier to ecological research because experts usually write keys for other experts, and not for general ecologists. The keys in these books are meant to be easy to use. Their usefulness depends very much on the illustrations, the preparation of which was assisted by a grant from the Natural Environment Research Council.

S.A.C.
R.H.L.D.

Acknowledgement

I should like to express my deep gratitude to Nathalie Yonow for her constant support and untiring, cheerful assistance throughout the production of this book.

Figures X.4-X.12, X.17 and X.21 have been redrawn from the book British Opisthobranch Molluscs by kind permission of Dr. T.E.Thompson.

P.J.H.

Contents

	Editors' preface	*page iii*
1	Introduction	1
2	Seaweed faunas	4
3	Herbivores and detritivores	11
4	Filter feeders	19
5	The fauna of kelp holdfasts	28
6	Predators	32
7	Identification	38
	Quick-Check key	45
	Main Keys	48
	I Sponges	48
	II Hydroids	49
	III Polychaetes	53
	IV Isopods	60
	V Gammaridean amphipods	62
	VI Caprellidean amphipods	70
	VII Pycnogonids	71
	VIII Lamellibranchs	74
	IX Prosobranchs	77
	X Opisthobranchs	81
	XI Bryozoans	86
	XII Ascidians	94
8	Techniques	97
	Some useful addresses	100
	References and further reading	101
	Index	106

Plates 1 - 8 are between pp. 48 and 49

1 Introduction

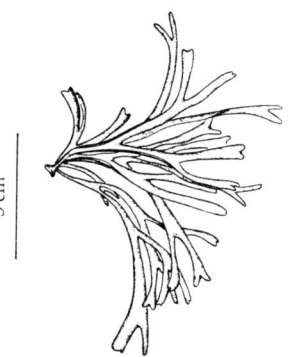

Fig. 1. *Pelvetia canaliculata* (L.) Done. et Thur.

detritus feeders eat particles derived from dead plants (or animals)

a sessile animal lives permanently attached to its substratum; e.g. a sponge

a sedentary animal moves little but is not immovably attached to its substratum; e.g. an anemone

epifauna: literally 'upon'; the assemblage of animals on living or non-living surfaces

a bryozoan

a hydroid

an ascidian or sea squirt

The rocky coasts of Britain are clothed by a broad belt of seaweeds occupying a zone extending from the high water mark of spring tides to about 20 metres below low water mark. The upper limit of this zone is set by the ability of the plants to withstand desiccation, and the lower limit by their ability to continue photosynthesis in increasingly dim and turbid waters. At the top of the shore seaweed cover is sparse and consists of tough, resistant species such as the Channelled Wrack *Pelvetia canaliculata* (fig. 1). Seawards, the essentially subtidal kelp forests gradually peter out into a scrub of small red algae. Between the tide marks dense growths of many different species of seaweed may flourish, distributed according to physical and biological constraints particular to each species.

The seaweed belt represents an immense biological resource and it is not surprising that diverse and often complex communities of animals exist to exploit it. Some algae are important food sources; others may act as sediment traps, providing a secondary source of food for detritus feeders; or are attractive by virtue of the films of micro-organisms which grow on their surfaces. However, the blanket of algae covering many rocky shores provides also a wide variety of microhabitats for numerous non-herbivorous species. It is important in supplying refuges for motile animals during low tide periods, but the immense surface area of algal tissue also provides living space for numerous sessile or non-sessile sedentary animals. On the middle and lower shore the constantly moving seaweeds may be heavily colonised by an epifauna of filter-feeding animals such as bryozoans,* hydroids,* ascidians* and sponges.

Seaweeds thus offer a range of resources for exploitation and the animals living on them may be utilising one or more of these resources. Some casual inhabitants make use of the alga for food, shelter or support but do not depend upon it. For example, Common Periwinkles *Littorina littorea* (fig. 2) graze most areas of the middle and upper shore on suitable coasts and commonly occur attached to algae. A variety of crabs, molluscs, worms, echinoderms and other organisms may be found on or under the bushy fucoid algae, but they will also occur, in

* Each group of organisms marked with an asterisk is illustrated by a standard silhouette in the margin as a reminder. The fuller classification appears in the Quick-Check Key on pages 45-47.

1 Introduction

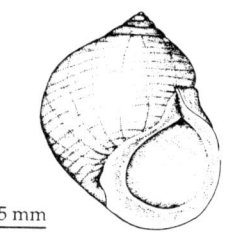

Fig. 2. *Littorina littorea* (L.).

a sea spider

fugitive/ephemeral: terms used to describe seaweeds which occupy a habitat only temporarily, until they are excluded by competition from other seaweeds, or by intense grazing by herbivorous animals

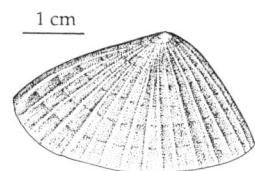

Fig. 3. *Patella vulgata* L.

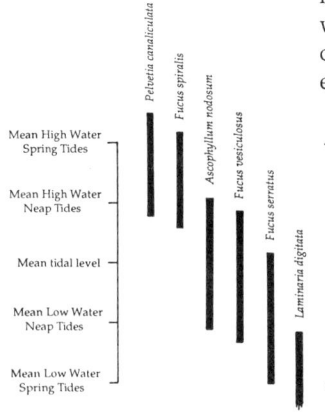

Fig. 4. Intertidal distribution of some common brown algae.

equally attractive areas of the shore, under stones and in crevices. However, a substantial number of invertebrate animals live either wholly or largely only on intertidal seaweeds and these are the subject of this handbook.

Three main types of algae may be recognised on most rocky shores. Brown algae (Phaeophyceae) usually comprise the bulk of the algal flora on sheltered to moderately exposed shores. These are particularly exemplified by the fucoid algae, such as *Pelvetia*, *Fucus* and *Ascophyllum*. They are typically large and bushy, with flat, often broad, regularly branching fronds which provide good support for encrusting organisms, and large surface areas for colonisation. The lower fringe of the intertidal zone is usually dominated by *Laminaria* species; the fronds of these kelps are less favourable to encrusting animals than those of the fucoids, largely because of their extreme flexibility, but the basal holdfast of the plant constitutes a species-rich microhabitat. Red algae (Rhodophyceae) tend to be small, and shrubby or filamentous. Their sessile epifauna is usually sparse, except on more exposed shores where *Fucus* species are scarce; the characteristic fauna of the *Fucus* species may then occur on the two red algae *Gigartina stellata* and *Chondrus crispus*. However, some red algae support high populations of small snails, sea spiders* (pycnogonids) and small crustaceans. Green algae (Chlorophyceae) are, on most shores, fugitive or ephemeral species, maintained at a low level by non-selective grazers such as limpets (*Patella*: fig. 3), or by competitive exclusion by red and brown algae. Nevertheless, where populations of such genera as *Cladophora* are able to develop they may support a number of interesting herbivores.

Intertidal algae are not distributed at random on any rocky shore. A basic zonation may be recognised (fig. 4) in which a number of spatially dominant species replace each other in sequence down the shore. The upper boundary to each species' distribution represents the coincidence of biological and physical factors which limit further upward expansion. The most important of these are the period of daily emersion, when the shore is uncovered by the receding tide (with consequences for photosynthesising efficiency), desiccation and, except in the case of the highest zoned species, grazing pressure and competitive exclusion. The lower limits to each seaweed zone are usually set by biological factors such as grazing and, most importantly, competition for space by the next zoned species. However, physiological factors may also be important as the photosynthesising ability of some midshore algae declines as the plant extends into deeper water.

These major seaweed zones are present on all rocky shores around Britain although they may vary locally according to other environmental features, the most important of which are shore aspect, profile and exposure. For example, on sheltered north-facing shores the upper boundaries of each zone may be moved upwards; gently sloping, sheltered shores will tend to have very broad midshore zones, while on steep shores the horizontal extent of all zones will be considerably reduced. 'Exposure' is sometimes difficult to quantify or describe, but essentially it relates to the degree to which a shore is affected by wave surge. As exposure increases the larger, bushy fucoids will tend to be replaced by shrubby red algae which, on even more exposed shores, will in turn give way to encrusting coralline algae. Zonation and exposure have been extensively studied on shores around Britain (Lewis, 1964).† An exposure scale has been formalised with a series of biological criteria for its assessment and comparison from locality to locality (Ballantine, 1961). For the purposes of comparing seaweed faunas between different shores it is important to realise that algal cover will vary from shore to shore; some experience in assessing exposure is useful and the works cited above include accounts of techniques and procedures for doing this. Further information will be found in Price and others (1980).

Animals living on seaweeds may be described as epiphytes, a usefully all-embracing term. It is used here to denote not only those sessile species which spend their lives attached to their algal host, but also specialised herbivorous species which are equally, though differently, wholly dependent upon an algal resource. Numerous other animals may occur on particular algae but are not primarily associated with them, although under certain conditions they may be strikingly abundant. Many of these have been included in the identification keys presented here. However, one other group of essentially non-epiphytic species has to be considered in studying algal epifaunas, namely the predatory species which may play important roles in the life cycles of the epiphytes. A number of the most important predators of sessile epiphytes have now been recognised, although many more doubtless await discovery by the keen investigator; some are known to be remarkably selective in their choice of prey and these have also been included in the identification keys. This book deals with epiphytes, as defined here, and their predators.

epiphyte: literally 'upon plants'; describes an organism, either plant or animal, which lives attached to plants

† References cited under the author's name in the text appear in full in the list of references on page 101.

2 Seaweed faunas

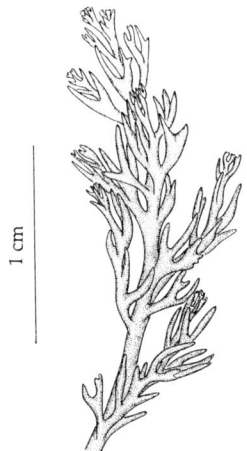

Fig. 5. *Cystoseira tamariscifolia* (Huds.) Papenf. : growing tip to show branching pattern.

The seaweed cover of a moderately sheltered rocky shore will be dominated by relatively few species, in strictly zoned sequence from the mean high water mark of spring tides (MHWS) to the extreme low water mark of spring tides (ELWS). By virtue of their abundance these species of algae will be most frequently utilised by epiphytic animals, although less common species such as the Peacock Weed *Cystoseira tamariscifolia* (fig. 5), which may be locally abundant on some southwestern shores, may also support substantial epiphytic communities. An early study of animal populations in seaweeds was made by Colman (1940), who counted individuals in weighed samples of weed from a shore transect at Church Reef, Wembury, near Plymouth, expressing his results as numbers of animals per 100 grams wet weight of each species of seaweed.

The figures (table 1) highlight the relative abundance of different groups of invertebrates in different species of weed. Such a survey is a useful preliminary to a study of seaweed faunas, but offers little insight into their ecology. The factors which influence the distributions of marine invertebrates have since been studied quite extensively (a good review is given by Meadows & Campbell, 1972) and it is now clear that observed

Table 1. *Mean populations of invertebrates in seven species of algae: individuals per 100 grams of weed*

	Pelvetia canaliculata	Fucus spiralis	Fucus vesiculosus	Ascophyllum nodosum and Polysiphonia lanosa	Fucus serratus	Gigartina stellata	Laminaria digitata holdfasts
Sponges	-	-	-	-	-	-	8.3
Coelenterates	-	-	-	3.0	0.2	216.0	31.0
Turbellarians and nemertines	-	0.3	5.3	63.6	0.5	2.8	40.2
Nematodes	3.4	3.2	3.3	76.1	21.2	16.5	247.8
Polychaetes	-	0.8	-	72.6	0.6	73.8	2056.0
Oligochaetes	0.2	3.2	0.3	39.2	0.2	-	9.7
Sipunculans	-	-	2.0	-	0.3	10.8	1.5
Ostracods	-	0.5	16.0	353.3	3.0	0.5	7.5
Copepods	-	25.8	221.0	272.2	178.1	1676.2	54.0
Barnacles	-	-	-	-	-	1.2	51.0
Tanaids	-	-	-	0.2	-	-	13.7
Isopods	15.2	0.5	4.3	30.0	8.1	32.2	6.3
Amphipods	15.4	46.3	1.3	48.1	4.3	83.8	125.3
Decapods	-	0.2	-	0.2	0.1	-	4.0
Sea spiders	-	-	-	0.2	-	-	3.7
Mites	-	3.2	135.7	222.4	75.6	758.8	7.3
Insects	0.8	1.3	2.3	58.3	2.0	-	-
Lamellibranchs	-	-	-	14.8	0.8	42.0	96.2
Gastropods	8.8	13.5	67.0	163.3	23.5	72.2	19.5
Bryozoans	-	-	-	0.2	6.1	158.2	73.5
Ascidians	-	-	-	-	-	-	7.8
Total	43.8	98.8	458.5	1417.7	324.6	3145.0	2864.3

From Colman (1940).

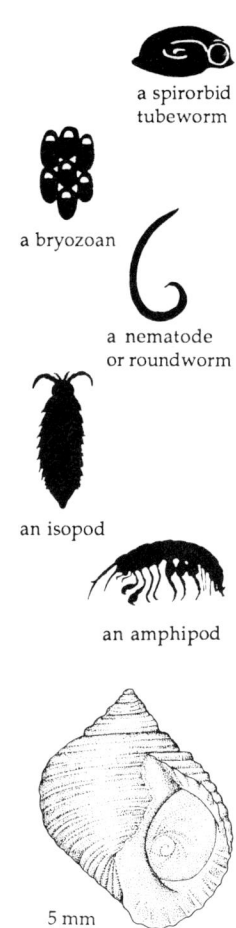

a spirorbid tubeworm

a bryozoan

a nematode or roundworm

an isopod

an amphipod

Fig. 6. *Littorina saxatilis* (Olivi).

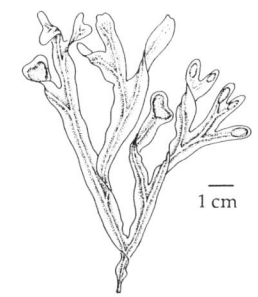

Fig. 7. *Fucus spiralis* L.

distributions result from behavioural patterns which help each species to find and remain in the habitat most suited to it. In species with narrow habitat requirements the distribution of the adult is often the result of a choice made by the prospecting larva. This is particularly true in the case of sessile or sedentary animals in which there is little or no opportunity for the adult to change its habitat. Larvae are dispersed during a brief free-swimming existence in the plankton, before settling on a suitable substratum and metamorphosing into the fixed adult form. For example, the sedentary spirorbid tubeworms* occur on a wide variety of algae, but each species of worm is usually limited to a narrow range of algal species, and so precise is their settlement behaviour that larvae often will not metamorphose successfully unless they locate their preferred algae. The same is true for some epiphytic bryozoans*, and also specialised herbivorous or detritus-feeding molluscs, which again may require contact with their usual substratum in order to complete metamorphosis. However, not all epiphytic animals depend upon larval selection of substratum in order to locate their preferred habitat. A number of strategies are employed and the contrast between these, even in equally specialised species, is often striking. Examples are introduced at appropriate places in the text. Table 1 comprises data for all animal species found in the Wembury transect. A substantial proportion of these will have been equally abundant under stones or in crevices elsewhere on the shore, and for many others, particularly the nematodes*, their biology is so little understood that their relationships with their algal habitat are still unknown. The keys presented here allow the identification of species considered to be primarily dependent upon algal habitats, and in this and the succeeding chapters the characteristics of epiphytic species will be examined and their adaptations to their algal hosts described.

The abundance and diversity of seaweed faunas increase down the shore. The high shore *Pelvetia canaliculata* supports few epiphytes, a fact attributable to the plant's habitat and the extreme environmental conditions it experiences. Most of the animal species recorded in *Pelvetia* clumps are seeking shelter during periods of low tide and consist mostly of wandering isopods* and amphipods*. The high shore periwinkle *Littorina saxatilis* (fig. 6) is often abundant on *Pelvetia*, the grooved fronds of the alga offering an alternative to its usual habitat in crevices or barnacle shells, and the snail shows no special attraction to the alga. No specialised epiphytes are associated with *Pelvetia*. *Fucus spiralis* (fig. 7), or Spiralled Wrack, has a slightly richer

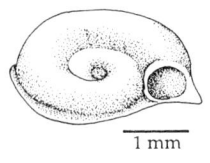

Fig. 8. *Spirorbis spirorbis* (L.).

fauna than *P. canaliculata*, although again there appear to be no species characteristically associated with it. That this is due to the harshness of its physical environment rather than to a lack of attractiveness on the part of the alga is suggested by the fact that larvae of some bryozoan species, when offered a choice of substrata to settle upon, will readily choose *F. spiralis* (table 2).

Table 2. *Percentage settlement of larvae of four bryozoans on some common seaweeds*

	Fucus serratus	Fucus spiralis	Fucus vesiculosus	Chondrus crispus	Gigartina stellata	Ascophyllum nodosum
Alcyonidium hirsutum	35	19	6	11	14	-
Alcyonidium gelatinosum	33	13	9	9	19	2
Flustrellidra hispida	37	23	15	3	13	3
Celleporella hyalina	9	6	1	25	14	3

Experimental data from Ryland (1959).

phenotype: the total characteristics of a plant or animal, determined by its genetic make-up, mediated or modified by environmental effects

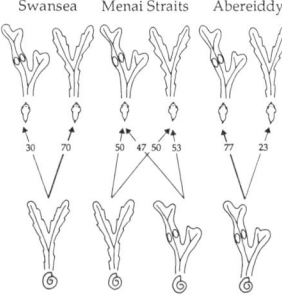

Fig. 9. Larval settlement of *Spirorbis spirorbis* from three different populations: percentage of larvae from each sample settling on *Fucus serratus* and *F. vesiculosus*. (Data from Knight-Jones and others, 1971.)

The absence of naturally occurring populations of these bryozoans on *F. spiralis* can only be explained by assuming that its high shore habitat is unsuitable for them. The Bladder Wrack *Fucus vesiculosus* (pl. 3.2) has a more diverse epifauna than *F. spiralis*. Few colonial organisms are associated with this alga but the tubeworm *Spirorbis spirorbis* (fig. 8) is common on it on some shores. *S. spirorbis* is most usually associated with the Serrated Wrack *Fucus serratus* (pl. 3.4), although a proportion of prospecting larvae will settle upon *F. vesiculosus*. However, at some localities not only does the preponderance of the population occur on *F. vesiculosus*, but larvae collected from that population, offered a choice of algal substrata under experimental conditions, will choose *F. vesiculosus* in preference to *F. serratus* (fig. 9), indicating phenotypic differences between populations of the worm (Knight-Jones and others, 1971). *Fucus vesiculosus* is also attractive to herbivorous isopods, such as *Idotea*, and to surface grazing snails, such as *Littorina obtusata*, although the occurrence of these is less the result of larval selection than of adult choice.

The middle shore algae *Fucus serratus* and *Ascophyllum nodosum* (pl. 3.3) are particularly favoured by epiphytes and substantial populations of sessile filter-feeding animals may occur on them. Most characteristic of *F. serratus* are the bryozoans *Alcyonidium hirsutum*, *A. gelatinosum*, *Flustrellidra hispida* and *Electra pilosa*. The first three are limited to the intertidal zone and live almost exclusively on *F. serratus*; on more exposed shores, where

a hydroid

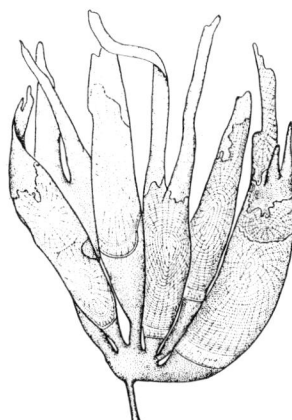

Fig. 10. Colonies of *Membranipora* on a kelp frond.

calcified: refers to tissues and structures reinforced by calcium carbonate

zooids: literally 'little animals'; the units (sometimes called polyps) of hydroid, bryozoan, entoproct and ascidian colonies, each functionally independent but an indivisible part of the whole colony

their preferred substratum is rare or absent, all three species will be found on *Gigartina stellata* and *Chondrus crispus*. *Electra pilosa* occurs commonly on a number of middle to lower shore algae, and also on some hard substrata, and extends well below ELWS. The tubeworm *Spirorbis spirorbis* is another characteristic epiphyte of *F. serratus*, as is the hydroid* *Dynamena pumila*, and both may achieve very dense coverage under suitable conditions. Specialised herbivores include the snails *Lacuna vincta* and *L. pallidula*, and the isopod *Idotea granulosa*. The last-named, however, may be found on several other lower shore algae.

The narrow, flexible, slimy fronds of the Knotted Wrack *Ascophyllum nodosum* offer a poor support for most encrusting animals. Bryozoans do not seem to like *Ascophyllum*; experiments indicate that it is unattractive to the larvae of most intertidal species, with the exception of *Bowerbankia imbricata*. This species develops flexible tufted colonies, rather than encrusting sheets, and may be locally common on *A. nodosum*, usually attached close to the air bladders of the plant. On very sheltered shores *Ascophyllum* may be co-dominant with *F. serratus* on the middle shore. On sheltered shores in Nova Scotia *Spirorbis spirorbis* is equally common on both algae. With increasing exposure, and concomitant water turbulence, the rigid calcareous tubes of *Spirorbis* will not survive the flexing movement of the *Ascophyllum* frond. Some research carried out in Canada (Doyle, 1975), and not so far repeated in Britain, suggests that long before the *Ascophyllum* population declines in the face of increasing exposure, the tubeworm population exhibits a change in larval behaviour, which leads to high selection of *F. serratus* and only minimal selection of *Ascophyllum*.

Below the *Fucus serratus* zone the sublittoral fringe is dominated by kelps, species of *Laminaria*. The fronds of these large algae are smooth, rather slimy and very flexible, and perhaps for these reasons bear few epiphytes. However, some organisms have managed to adapt to this habitat, including several very specialised and successful species. The bryozoan *Membranipora membranacea* (fig. 10) is one; it is widespread along the Atlantic coasts of Europe and North America, and although small colonies are occasionally found on other seaweeds it is essentially adapted for living on the fronds of various species of *Laminaria*. *Membranipora* grows fast, forming extensive, white lacy colonies; they are very lightly calcified compared with those of many other bryozoans and small cracks in the vertical walls of the individual zooids render the colony as flexible as its substratum.

Another characteristic epiphyte of *Laminaria* is the

meristem: the part of a plant's tissue responsible for its growth. In seaweeds this may occur at the tip of the frond or between the frond and the stipe

stipe: the stalk of an alga, that part between the holdfast and the frond

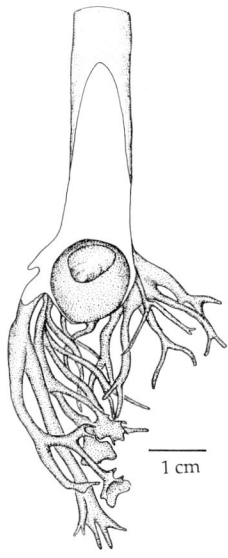

Fig. 11. A kelp holdfast cut to show the basal cavity containing a large *Patina*. (After Kain & Svensden, 1969.)

Fig. 12. Percentage occurrence of *Patina* cavities in *L. hyperborea* at Isle of Man. (After Kain & Svensden, 1969.)

Blue-rayed Limpet *Patina pellucida* (pl. 6.4). This small slipper-shaped limpet has a translucent, horn-coloured shell with striking, peacock-blue rays radiating from its apex to the broad anterior end. These limpets occur either singly or in small groups close to the basal region of the frond, often with their anterior ends orientated towards the stipe. They are rather small, rarely more than 6 millimetres in length, but larger individuals with duller shells may be found on the stipe of the alga, and very large specimens (more than 1 centimetre long) with dull, chalky shells, often denoted 'variety *laevis*', may occur in the holdfast. *P. pellucida* actually feeds on *Laminaria* tissue. It is an annual species. Larvae settle from the plankton in May and small limpets are abundant on the kelp fronds in June and July; the pits in the surface of the algal tissue beneath each limpet attest to voracious feeding. During the winter all kelp species lose or shed their fronds, and many *Patina* probably die as a result. In October and November the limpets begin a migration towards the basal regions of the frond and enough will survive the winter on the meristem or stipe of the alga. These will breed in the spring and then die before their offspring settle. However, a small number will continue to migrate down the stipe of the kelp, enter the holdfast and proceed to feed there, excavating in the process a substantial cavity at the base of the stipe (fig. 11). The lifespan of these individuals appears to be much more than a year, but for how long they live is unknown. Further, their reproductive contribution to the population is also unknown, and may perhaps be quite significant. It is clear, though, that 'variety *laevis*' has a significant impact on the ecology of kelp populations. In a *L. hyperborea* (Gunn.) Fosl. bed off the Isle of Man holdfast infestation by *Patina*, at 1 metre below ELWS, increased from less than 10% in 2-year-old plants to 67% in 7-year-old plants (Kain & Svensden, 1969). Infestation rates were lower 11 metres below ELWS (fig. 12)

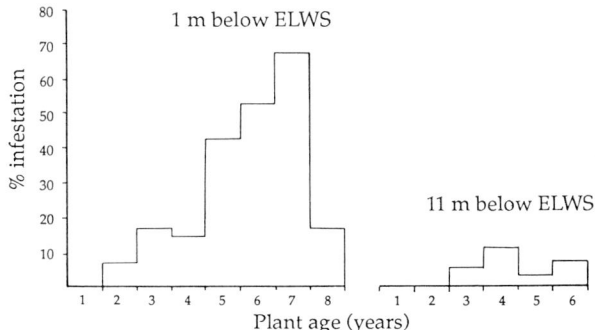

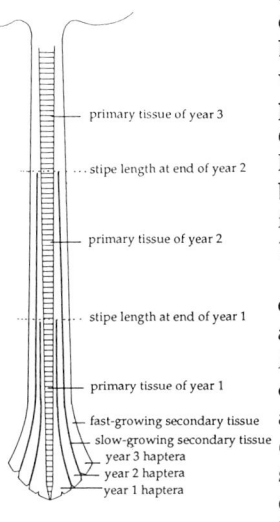

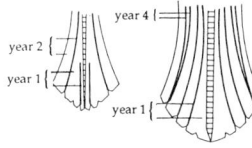

Fig. 13. Ageing *Laminaria hyperborea*: median longitudinal sections through stipes. Note that each yearly increment comprises primary tissue and a layer each of fast-growing and slow-growing secondary tissue. (After Kain, 1963.)

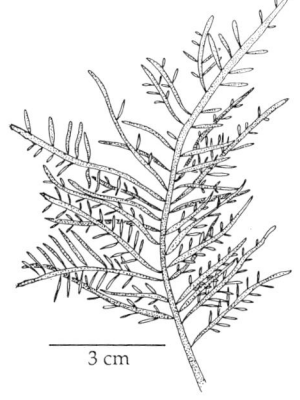

Fig. 14. *Gelidium sesquipedale* (Clem.) Born. et Thur.

and at 20 metres *Patina* appeared to be absent. Ebling and others (1948) found that on *Saccorhiza polyschides* (Lightf.) Batt. numbers of *Patina pellucida* were at a maximum when water currents passing over the frond averaged 1.3 metres per second and declined at greater or lesser velocities. The ecology of *P. pellucida* still poses problems; curiously, in Norwegian populations of *L. hyperborea* the limpet seems to be strictly an annual species and no individuals can be found in the plants' holdfasts (Kain & Svensden, 1969; Vahl, 1971).

The stipes of all *Laminaria* species are often thickly covered with red algae, erect bryozoans, hydroids, sponges and ascidians, and the red algae themselves, particularly *Palmaria* (pl. 4.6) and *Delesseria* (pl. 4.7), may support a diverse epifauna. There have been numerous descriptive accounts of the fauna and flora of *Laminaria* stipes but the ecology of this often rich habitat has still to be properly studied. By contrast, the ecology of the kelp holdfast community has been intensively studied. Kelps are long-lived plants; the stipe thickens by annual increments of tissue (fig. 13), and as the plant increases in size and weight the holdfast develops successive whorls of haptera, short rootlets which broaden its basal attachment area and anchor the plant more firmly. Thus, with increasing age the holdfast develops a more complex form, with greater surface area for the attachment of sessile animals and increasing volume between the haptera for sedentary organisms. The community of a large holdfast may comprise a very large number of species, and several thousand individuals. This topic is covered in detail in chapter 5.

Red algae tend to be distributed from mean tidal level (MTL) to the lower shore and are rarely abundant on fucoid-dominated shores. However, on near-vertical surfaces, in conditions of moderate exposure, sizeable populations may develop and these will harbour a rich epifauna. *Gigartina stellata* (pl. 4.4) and *Chondrus crispus* (pl. 4.5) may provide alternative substrata for the *Fucus serratus* community where *Fucus* is sparse or absent, although neither has a unique epifauna. Other species, such as *Lomentaria articulata* (pl. 4.1), *Laurencia pinnatifida* (pl. 4.3), *Corallina officinalis* (pl. 4.2) and *Gelidium sesquipedale* (fig. 14), with more slender fronds and a less open structure than the two species above, support a very interesting fauna. The delicate, filamentous morphology of these algae constitutes an efficient sediment trap and quantities of fine-particled detritus are passively filtered from the sea and accumulated through the spring and summer months. This sediment is an important food source for the newly settled juveniles of many small snails, such as *Rissoa, Tricolia pullus* and *Barleeia*

a sea spider

unifasciata, and enormous populations may build up on the plants. Adults may proceed to feed on the algal tissue, or continue to subsist on detritus; some species feed on the sessile diatoms (single-celled algae) which also flourish on these small algae. The plants are too delicate to support encrusting, filter-feeding animals, and their tendency to form dense, detritus-laden mats is probably another factor discouraging sessile animals. Small sea spiders*, whose feeding preferences are not fully known, are also frequently common in these communities.

A reasonably detailed field study of epiphytic animals may be accomplished with reference to the commoner, and spatially dominant, seaweed species mentioned in this chapter. Illustrations of all these have been provided in this handbook. However, a comprehensive survey may demand that all seaweed species encountered be accurately identified. A key to all intertidal seaweeds is beyond the scope of this work, but good textbooks and handbooks to British seaweeds are available. Dickinson (1963) and Hiscock (1979, 1986) will be particularly useful.

3 Herbivores and detritivores

Herbivorous animals, particularly the molluscs, have a profound impact on the algal flora of rocky shores. Limpets (*Patella*) and Common Periwinkles (*Littorina littorea*) are among the most significant grazers on British shores, and below mean tide level (MTL) Flat Periwinkles (*L. obtusata, L. mariae*) and top shells (*Gibbula*) may be equally important. Together with numerous other, less common species these herbivores play a major part in the structuring of intertidal algal communities. Interesting field experiments may be devised to demonstrate the effects of major herbivores: for example, exclusion by cages or fences of all, or selected, species of grazers from limited areas of shore at different tidal levels has often been employed to investigate interactions between all organisms, and shows how grazing most often acts to maintain species diversity. More simply, qualitative and quantitative changes in algal cover and zonation may be recorded following removal of all grazers over a broad shore transect; observations should be continued through the period of natural recolonisation by grazers, and compared with those from undisturbed control areas. This kind of manipulative study is a well-established technique in intertidal ecology (see Price and others, 1980; Barnes & Hughes, 1982) and still offers considerable potential for further work (see Lewis, 1980).

Fig. 15. *Halidrys siliquosa* (L.) Lyngb.

Although these dominant grazers may have food preferences in terms of plant size, and may have their greatest effect at or between certain tidal levels, they are essentially non-selective and the survival of their populations is not dependent upon one or few algal species. In this sense they do not fall within the terms of reference of this handbook. However, there are a number of important herbivorous species whose life cycles appear to be closely attuned to those of their algal substrata, and whose survival appears to be completely dependent upon that of their algal habitat. These would seem to be properly termed epiphytic, and some particularly interesting examples are discussed in this chapter.

Lacuna pallidula (pl. 7.4) and *L. vincta* (pl. 7.1), two small grazing snails, were the subjects of a detailed study by Smith (1973), who investigated their ecology on the coast of Durham. On most shores *L. pallidula* is practically confined to *Fucus serratus*, although a small proportion of each population may occur on adjacent *F. vesiculosus*. *L. vincta* has a more catholic range of substrata; it occurs on *Fucus serratus, Laminaria, Halidrys siliquosa* (fig. 15), *Palmaria palmata* and a range of smaller red algae, but on any one shore the bulk of the population will probably be found to occur on just two or three species of seaweed. Both species feed directly on their algal host, are significant consumers of algal tissue, and appear to display a close synchrony between their life cycles and those of their food plants (fig.16).

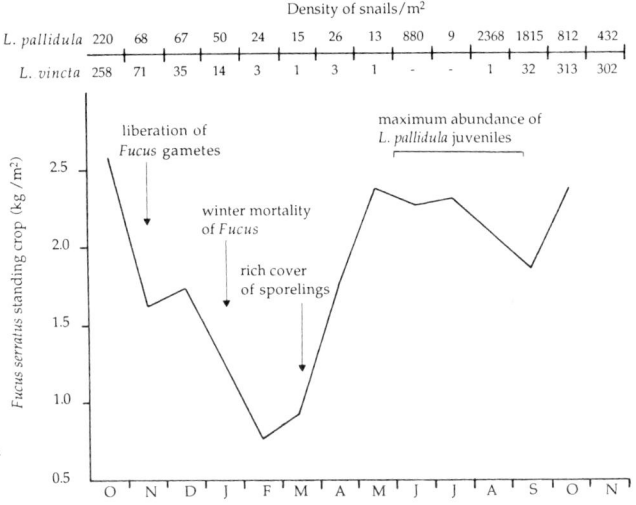

Fig. 16. Relationships between *F. serratus* crop and *Lacuna* species at Whitburn, Co. Durham. (Data from Smith, 1973.)

direct development: in certain molluscs, a reproductive mode in which development of the fertilised egg, through larval stages to juvenile, takes place within an egg capsule. On hatching, a small version of the adult is released

veliger: the swimming planktonic molluscan larva. Locomotion is achieved by two large, ciliated body lobes - the velum

L. pallidula commences reproduction in November; spawning begins in January and peaks in April, and the distinctive spawn masses are laid directly on the *Fucus serratus* plant, either close to the new tissue at the tips of the plant, or on very young plants. The embryos undergo development to an advanced stage within the egg capsule, and hatch as crawling juveniles, a process referred to as 'direct development'. Newly hatched juveniles may be found in groups around the jelly of the egg capsules. *L. pallidula* is an annual species with a long breeding period; juveniles are recruited into the population from April to July, which coincides with the period of vigorous growth of *F. serratus* sporelings, and population densities peak in late summer. *L. vincta* is also an annual species but has an even more protracted breeding season; spawning begins in January and reaches a peak in June. Egg capsules are laid directly onto the food alga of the adult but, in contrast to *L. pallidula*, the eggs hatch into plankton-feeding veliger larvae which spend some weeks in the plankton before settling. Newly settled juveniles may be found from June to October, with numbers reaching a maximum in September.

The breeding cycles of both *L. pallidula* and *L. vincta* tend to be attuned to those of their host algae; on the south Devon coast *F. serratus* reproduces earlier in the year than the Durham populations, the summer bloom of sporelings consequently occurs earlier in the season, and *L. pallidula* and *L. vincta* breed equally earlier. In both species adult snails die after breeding; in *L. pallidula* generations may overlap during one month of the summer, and in *L. vincta* during perhaps two months. However, both species have very long breeding periods - five to six months - and size/frequency histograms of samples may reveal several distinct classes, or 'cohorts', representing successive age groups. Smith (1973) has suggested that an extended breeding period is a necessity in an animal with an annual life cycle, reducing the probability of local extinction through catastrophic environmental change.

In one aspect of their life cycles the two species show an interesting contrast, reflecting two different adaptive strategies. *L. pallidula* ensures that its offspring find themselves in their optimal habitat through direct development from egg capsules deposited on the food alga. *L. vincta* larvae complete their development in the plankton and, perhaps because the settling veligers are unable to differentiate between different species of algae, the adults occur on a number of algal species. Habitat selection at settlement is mostly a feature of species with short swimming larval stages; settlement after a relatively prolonged planktonic existence is a hazardous undertaking because many larvae will have been carried far away from suitable habitats. Adaptation to a limited range of microhabitats seems to have been an expensive process for both species of *Lacuna*, perhaps because their larvae lack behavioural mechanisms which would permit more precise substratum selection before metamorphosis. Direct development is an energetically expensive undertaking, demanding the production of enough yolk to nourish the embryo throughout its development. Grahame (1977) measured the reproductive output of captive populations of the two *Lacuna* species and showed that *L. pallidula* produced a mean of 1365 eggs per female. However, *L. vincta*, coupling a more inefficient larval cycle with a narrow range of optimal adult habitats, produced an astonishing mean of 53 432 eggs per female. Although the unit cost of plankton-feeding larvae is far less than that of yolk-feeding larvae, the vast numbers of eggs produced by *L. vincta* required that a substantially greater proportion of the snail's energy budget be directed towards reproduction.

Grazing by *Lacuna* has severe effects on algal populations, particularly of *Fucus serratus*. *L. vincta*, perhaps as a result of the species' adaptive strategy, may achieve plague proportions in some years, when the algal substrata may be severely defoliated. A field study on the coast of New Brunswick, Canada, quantified the effects of a plague of *L. vincta* on its preferred substratum, *Fucus edentatus* De La Pyl (Thomas & Page, 1983). The alga achieved a mean production of 61 grams dry weight per square metre per day during the summer, and it was estimated that *L. vincta* removed 79% of this production. Weight loss of algal tissue was dramatic (table 3), heavily grazed plants were reduced practically to the midribs and stipe, and in response the plants stopped growing at their tips and grew by basal thickening.

Table 3. *Effect of grazing by* Lacuna vincta *on* Fucus edentatus *De la Pyl populations in New Brunswick, Canada*

	Lightly grazed		Heavily grazed	
	Mean dry wt (mg)	Mean length (mm)	Mean dry wt (mg)	Mean length (mm)
June	191.73	93.36	85.15	73.24
July	484.67	142.72	160.29	84.84
August	882.47	194.52	186.25	101.00

From Thomas & Page (1983).

The ecology of the two species of *Lacuna* in Britain would repay further investigation, as would behavioural studies on adult snails in relation to substratum selection, and the responses of *L. pallidula* populations to changes in *Fucus serratus* populations. Southgate (1982) has recently conducted a comparative study of populations of both species in red algal turfs on the west coast of Ireland. Although *L. pallidula* juveniles occurred frequently on red algae, adults were found to be restricted to *Fucus serratus*. At Wembury, south Devon, *L. vincta* occurs abundantly on small red algae together with two other small snails, *Tricolia pullus* (pl. 7.6) and *Cerithiopsis tubercularis* (pl. 7.3). The ecology of these three species was studied by Fretter & Manly (1977). *Tricolia pullus* juveniles feed on the algae on which they settle from the plankton, and on sessile diatoms growing on the algae. Adults also feed on algal tissue, and on the fine detritus trapped and retained by the alga. *Cerithiopsis* has a special habitat on sponges such as *Hymeniacidon* and *Halichondria*, on which it feeds and lays its eggs. These sponges occur often in association with certain red algae, closely intergrown with the holdfasts, and are the primary substrata of the adult snails. Newly settled juveniles, as small as 0.32 millimetres long, were found on those algae most often associated with sponges. The numbers of snails found by Fretter & Manly (1977) are impressive (table 4) and give a good idea of the richness of these microhabitats.

Table 4. *Mean occurrence of three gastropod species from June to September at Wembury, expressed as numbers per gram dry weight of alga*

	Lacuna vincta	Tricolia pullus	Cerithiopsis tubercularis
Lomentaria articulata (Huds.) Lyngb.	382.5	36.0	22.6
Laurencia pinnatifida (Huds.) Lamour.	156.7	14.0	4.3
Gigartina stellata (Stackh.) Batt.	77.3	7.8	-
Corallina species	67.0	12.7	19.0
Gelidium sesquipedale (Clem.) Born. & Thur.	57.0	7.0	2.3
Ceramium species	42.0	2.2	1.8
Chondrus crispus Stackh.	19.5	2.0	-
Cladophora rupestris (L.) Kutz	1.25	-	-
Plumaria elegans (Bonnem.) Schm.	0.9	8.0	-
Fucus serratus L.	0.8	-	-
Callithamnion species	0.01	-	-

From Fretter & Manly (1977).

All three species have annual life cycles, with long breeding seasons and a short overlap in generations. Size/frequency histograms again revealed successive cohorts, and also showed that the juveniles occurred on the algal species favoured by the parent snails. The suitability of each alga depended upon its morphology and its ability to trap and retain sediment. The most suitable were dense turf-forming species with richly branched fronds 30-160 millimetres long which provided maximum surface area for diatom growth, and retained substantial quantities of silt, particularly in the holdfast. A uniformly distributed flora of sessile diatoms, with maximum species diversity, seemed to be more attractive to young snails than irregular, though dense, masses with lower diversity. Overall, the most favoured algae were those which provided stable, protected microhabitats for their snail fauna, and presumably for numerous other organisms as well. That this was by virtue of their morphology, rather than through intrinsic properties of the algae themselves, was demonstrated by the fact that these microhabitats were stable only within certain ranges of exposure. As exposure increased beyond these limits, and the microhabitats disintegrated, the incidence of all three species of snail declined sharply.

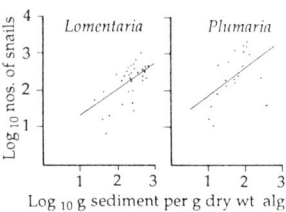

Fig. 17. Relationships between numbers of *Rissoa parva* and sediment load in two red algae. (From Wigham, 1975.)

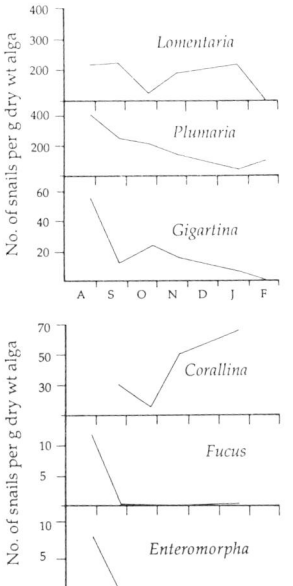

Fig. 18. Seasonal variation in numbers of *Rissoa parva* on different species of seaweed. (From Wigham, 1975.)

veliger: the swimming planktonic molluscan larva. Locomotion is achieved by two large, ciliated body lobes - the velum

The red alga *Laurencia pinnatifida* was the second most important substratum for *Lacuna vincta* at Wembury, yet it had one serious disadvantage for the snail. During July and August the alga underwent drastic defoliation when old fronds were shed, resulting in the loss of large numbers of *L. vincta*, and a corresponding sharp decline in mean overall population density. Another detritus-feeding snail, *Rissoa parva* (pl. 7.2), was also common on the algae surveyed by Fretter & Manly (1977) but achieved maximum population densities during July and August. Like *Lacuna*, *R. parva* was frequent on *Lomentaria articulata*, but was also common on the more open-structured species such as *Ceramium*, *Cladophora* and *Callithamnion*, which were generally unsuitable for *Lacuna* and *Tricolia*. *Rissoa* possesses an adhesive gland on the foot enabling it to attach itself firmly to its algal host, and so was less dependent upon a dense algal structure. The biology of *Rissoa parva* was studied in the same area of south Devon by Wigham (1975), who demonstrated clearly the correlation between the abundance of the snails on favoured algae, and the silt load of the algae (fig. 17). *R. parva* is another annual species, with a long breeding period; larval settlement occurred in most months at Wembury, with peaks in May, August and November. However, the population density of the snail did not follow such a simple pattern but varied according to which alga was sampled (fig. 18). On some weeds it declined during the winter months, on others it remained high, and on *Lomentaria* it actually increased. Some algae were less resistant to winter storms, while others simply offered less protection for *Rissoa*. The decline on *Plumaria* and the increase on *Lomentaria* were attributed to the migration of snails from 'summer pastures' on the former to winter quarters on *Lomentaria*.

Fretter & Manly (1977), remarking on the fact that juvenile snails occurred on the same algal substrata as the adults, postulated that this might indicate habitat selection by settling veliger larvae. The stimulus to settlement could have been provided by attractant properties of the algae, of the diatom flora, the sediment, or even of the adult snails themselves. However, Wigham (1975) conducted controlled experiments with freshly caught veligers of *R. parva*, offering them a range of substrata for settlement, and was unable to find any evidence of larval selection of habitat. Most interestingly, he pointed out that the size of the settling veligers was within the size range of silt particles trapped by the most favoured algae, and suggested that the concentration of veligers on these weeds was simply a passive process. Under these circumstances adaptation of these species of snail to their particular algal substrata would be favoured and would not depend upon behavioural adaptations of the larval stage.

Barleeia unifasciata (pl. 7.5) is yet another small detritus-feeding gastropod, found in red algal turfs on western and southwestern shores (Southgate, 1982). Like the species discussed above, it occurs predominantly in algae such as *Laurencia*, *Lomentaria* and *Corallina*, which develop tight structures, ensuring a good food supply and offering maximum protection against wave crash (fig. 19).

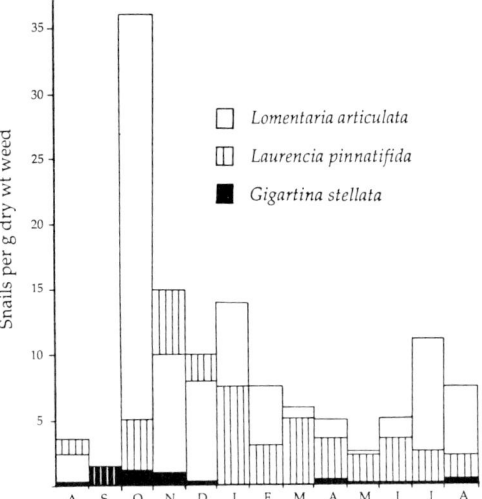

Fig. 19. Seasonal occurrence of *Barleeia unifasciata* (Montagu) on three species of red algae. (Data from Southgate, 1982.)

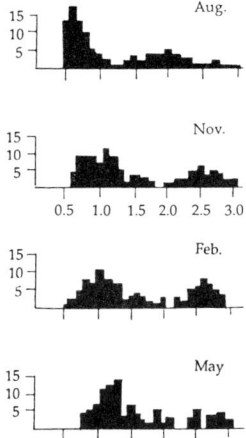

Fig. 20. Population structure in *Barleeia* in four monthly samples. (Data from Southgate, 1981.)

Unlike the preceding species it does not have a planktonic larva but rather, like *L. pallidula*, ensures the concentration of juveniles on suitable algae through the production of sessile egg capsules and direct development of larvae. The population studied by Southgate (1982) appeared to be breeding continuously, with juveniles, 0.5 millimetres long, present throughout the year (fig. 20). Recruitment was highest between June and September and maximum population densities occurred in the autumn.

Comparative studies of the faunas of red algal turfs present certain problems. The algal composition of the turf may vary horizontally and vertically on a shore and, as shown above, different algal morphologies favour different animal species. Further, organisms may be stratified within the turf, with some species occurring predominantly on the fronds of the algae, and others in the basal holdfasts. As the ratio of holdfast to frond varies seasonally, quantification of samples in terms of dry weight of whole plants is perhaps not a good basis for comparison between different turfs, as each algal species has a characteristic annual growth cycle. Myers & Southgate (1980) tackled these problems by employing artificial substrata to study turf faunas on different shores. They cleared patches of rock within areas of turf, blow-lamped them dry and attached flat or spherical nylon mesh saucepan scourers to the bare substrata, using epoxy resin glue or steel masonry nails. The animal communities which developed in the pan scourers were very similar to those in adjacent algal turfs. Although neither artificial substratum was directly comparable in structure or community to any one algal species, the faunas they developed fell within the range of associations found in natural mixed turfs (fig. 21), and both were found to be convenient and comparable sampling units.

Finally, one of the most interesting groups of epiphytes found on rocky shores is the sacoglossan sea slugs. These are specialised herbivores adapted for suctorial feeding and occur principally on coenocytic algae, in which the lack of cell walls enables the sacoglossan to gain the maximum return each time it pierces the outer wall of the plant. *Limapontia capitata* (pl. 5.1) is found almost entirely on species of the filamentous green alga *Cladophora*, in middle and upper shore rock pools. It is not known how adult *Limapontia* recognise or select their food plant. They do not seem to respond to chemical clues (Jensen, 1975) and show no preference in selection experiments. Algal morphology may be important, or perhaps the settlement behaviour of the larval stage concentrates the population in its high shore habitat and obviates the need for further precision. However, although *L. capitata* may be extremely abundant on some shores in early summer practically nothing is known of its ecology. Similar questions may be posed regarding the larger, and more beautiful, *Elysia viridis* (pl. 5.2, 5.3), which may be found on southern and western shores on the fleshy, coenocytic green alga *Codium*. *Elysia* is a very specialised herbivore whose diet is supplemented by the chloroplasts of its host plant. These are ingested by the slug and retained within cells of its gut diverticula where they continue to photosynthesise for some weeks. The carbon fixed by the chloroplasts evidently makes a significant contribution to the slug's nutrition. Although *Elysia* is able to survive for at least a month when fed on other seaweeds, it will only maintain its body weight when fed on *Codium* (Gallop and others, 1980). Despite this dependence, *E. viridis* seems to be incapable of selecting its substratum from among an array of seaweeds. Observations suggest that it will climb any seaweed it encounters under experimental conditions (Gallop and others, 1980), although it will only remain on *Codium*. It is possible that the slug continues to crawl until it is able to feed successfully.

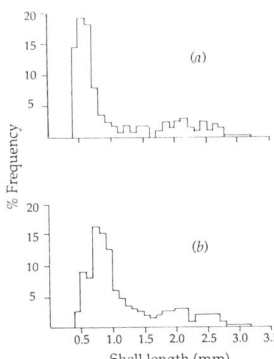

Fig. 21. Population structure of *Rissoa parva* on *Lomentaria articulata* (a) and artificial substrata (b) in June. (From Myers & Southgate, 1980.)

suctorial: with mouthparts specialised for sucking

coenocytic: describes algal tissue consisting of multinucleate cytoplasm not divided into separate cells

4 Filter feeders

a hydroid

a bryozoan

a spirorbid tubeworm

an ascidian or sea squirt

Assemblages of sessile filter-feeding animals become increasingly prevalent on seaweeds below mean tidal level, and on the lower reaches of well-sheltered shores they may be especially abundant. Many species of algae will support sessile epiphytes but the most dense and diverse communities are found on *Fucus serratus*. They include sponges, hydroids*, bryozoans*, tubeworms* and ascidians*. The sponges, such as *Scypha compressa* (pl. 1.2) and *S. ciliata* (pl. 1.1), and certain ascidians may be sufficiently common to be significant and successful competitors for space in an often crowded microhabitat. However, their occurrence and abundance are usually more directly related to environmental factors than to the inherent qualities of their substratum, and they are not here regarded as primarily epiphytic. In other habitats they may be equally common on non-algal substrata. Conversely, the spirorbid tubeworms, a number of bryozoan species, and the hydroid *Dynamena pumila* are truly characteristic of the *F. serratus* epifauna and are closely dependent upon their algal substrata.

Fucus serratus is particularly favoured as a support by sessile filter feeders. Through most categories of exposure it is the most abundant seaweed on the lower shore and its relatively broad, semi-rigid fronds, and dense, bushy habit provide an ideal habitat for many encrusting animals. However, other seaweeds may also bear substantial populations of sessile animals. On well-sheltered shores *Ascophyllum nodosum* may be particularly common. Although its narrower, more flexible fronds offer less secure support, it may provide an alternative habitat for some of the *F. serratus* epiphytes. As the abundance of *F. serratus* declines with increasing exposure three of its characteristic epiphytes - the tubeworm *S. spirorbis*, and the bryozoans *Alcyonidium hirsutum* and *Flustrellidra hispida* - will occur on *Gigartina stellata* and *Chondrus crispus*, although observations suggest that as exposure increases still further only *Flustrellidra* is able to persist on these small red algae. The smooth fronds of the kelps *Laminaria digitata* (Huds.) Lamour. and *L. hyperborea* are perhaps too flexible to support a wide array of sessile epiphytes and only the specially adapted bryozoan *Membranipora membranacea* is commonly present. However, in particularly sheltered conditions species of *Spirorbis*, various bryozoans, hydroids and ascidians may be able to maintain populations on *Laminaria* fronds, although none of the species is in any way

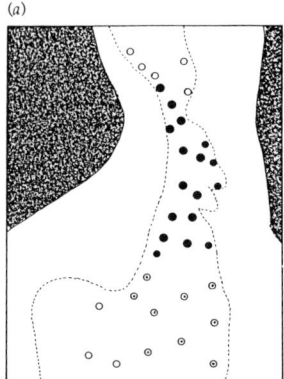

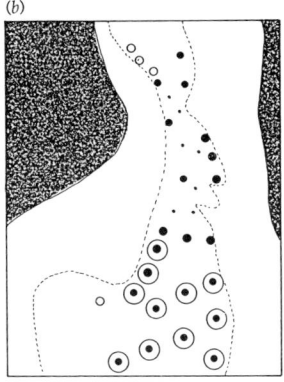

Fig. 22. Distribution of two species of *Laminaria* and their bryozoan epiphytes in a tidal rapids system. (*a*) Filled circles, *L. digitata*; open circles, *L. saccharina*; circle + spot, *L. saccharina* predominant over *L. digitata*. (*b*) Filled circles, *Membranipora membranacea*; open circles, *Celleporella hyalina* + *Callopora lineata*; circle + spot, abundant *Celleporella* + *Callopora*, with frequent *Membranipora*. Note: the dotted line = mean low water mark; current speed increases through narrows. (Modified from Ryland & Nelson-Smith, 1975.)

adapted for life as a *Laminaria* epiphyte. In very sheltered waters fronds of the subtidal Sugar Kelp *L. saccharina* (L.) Lamour. may be densely settled by encrusting organisms. Ryland & Nelson-Smith (1975) studied the distribution and abundance of *Laminaria* epiphytes in the tidal rapids system of a very sheltered bay in western Ireland and showed particularly well how the diversity and density of the epifauna were related to current flow (fig. 22). Many species of seaweed, then, under particular environmental conditions, will support a variety of sessile, filter-feeding animals. However, of these seaweeds only *Fucus serratus* has a relatively constant and characteristic epifauna, including a number of species which appear to be particularly adapted for life on this substratum.

In sessile animals the distribution of the adult stages is a direct consequence of the settlement behaviour of the larval stages. Once the larva has attached itself and metamorphosed it is rarely able to change its position. The factors influencing habitat selection by marine invertebrate larvae are numerous. Responses to light, gravity and current are particularly important, and perhaps for some species are sufficient to concentrate settling larvae within a range of suitable habitats. For some species, such as small herbivorous or detritivorous snails, blanket settlement of larvae within broad limits of environmental suitability may be all that is necessary to ensure continual recruitment to the population. For sessile species, particularly those with narrowly defined habitat requirements, this would be a dangerous proceeding and in those species which have been studied it is apparent that the choice of the prospecting larva is narrowed by responses to very specific stimuli. For example, spirorbid larvae will invariably choose to settle in pits, grooves or concavities on a surface (fig. 23), where they perhaps gain some protection during the critical period when the newly metamorphosed larva is establishing itself firmly in its new habitat. The larvae of many marine invertebrates are attracted to surfaces filmed with microorganisms (Meadows & Williams, 1963) and are often able to distinguish between films composed of suitable or unsuitable microfloras. In the case of most sessile epiphytes experiments have shown repeatedly that the algal substratum exerts the greatest influence on the settling larva.

Larvae of *Spirorbis spirorbis*, freshly collected from brooding adults, when offered a choice of algal substrata will preferentially settle upon *Fucus serratus*. This behaviour of *Spirorbis* larvae has been extensively researched and it was demonstrated by Williams (1964) that the attraction to the larvae lay in some inherent quality of the algal

4 Filter feeders

metamorphosis: a stage in the life history of many animals when the larva undergoes rapid and profound change to give rise to the adult form

substratum. By soaking fresh, coarsely chopped pieces of *F. serratus* in seawater Williams prepared an extract of the seaweed. Inert settlement panels, which had been allowed to develop microbial films by immersion for several days in seawater, were then soaked for 12 hours in the *Fucus* extract. Tested against combinations of unconditioned, filmed and unfilmed surfaces in clean seawater, the larvae settled overwhelmingly on the filmed, *Fucus*-conditioned panels (table 5).

Table 5. *Settlement of* Spirorbis spirorbis *larvae on tufnol panels*

Experiment number	Unfilmed panels	Filmed panels	Filmed panels + *Fucus* extract	Total larvae settled
1	0	7	390	397
2	0	0	5	5
3	0	0	103	103
4	46	111	3359	3516
5	0	142	1106	1248
6	0	50	208	258
7	1	5	25	31
8	2	12	86	100
9	0	9	287	296
10	1	1	50	52
11	0	1	32	33
12	0	11	28	39
13	2	2	30	34
14	0	4	698	702
15	4	1	19	24
16	17	6	49	72

From Williams (1964).

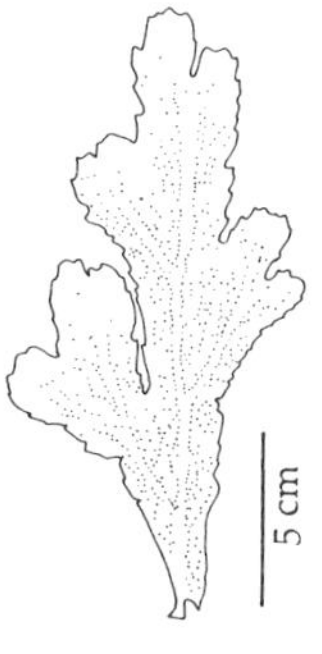

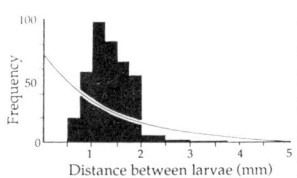

Fig. 23. A dense settlement of *Spirorbis spirorbis* larvae on a piece of *Fucus serratus*.

Fig. 24. Distance between *Spirorbis* larvae settled along the groove flanking the midrib of *Fucus serratus*. Histogram shows actual measured distances, compared with a curve for a random distribution of distances.

Subsequently, experiments with other epiphytic species of *Spirorbis*, and with *S. spirorbis* populations living on *F. vesiculosus* (see fig. 9), have revealed the same behaviour; the larvae consistently selected their usual algal substratum when offered a choice. The larvae of *S. spirorbis* also display a well-marked gregariousness, which in experiments may sometimes partly mask the strength of their attraction to the algal substratum. Such behaviour is of obvious importance in sessile, sexually reproducing organisms; larvae will invariably choose to settle close to adults or to previously settled juveniles and dense aggregations develop in natural populations. The dangers of overcrowding are mitigated by the larvae spacing themselves out at settlement so that even at the highest densities each larva will allow itself sufficient space to achieve a minimum diameter of about 1 millimetre (fig. 24). *Spirorbis spirorbis* has an annual life cycle and a maximum individual lifespan of up to 16 months. As in the case of the small snails discussed in the previous chapter, the tubeworm has an extended breeding period. In a

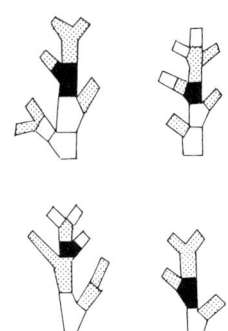

Fig. 25. Settlement density of *Alcyonidium* larvae on fronds of *Fucus serratus*. Black, high density; stipple, intermediate density; white, low density.

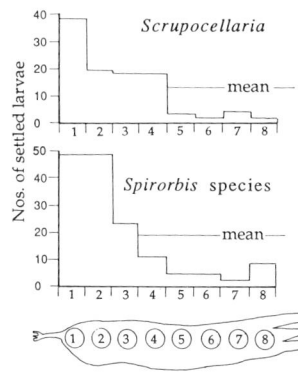

Fig. 26. Settlement of *Scrupocellaria* and *Spirorbis* species larvae on discs cut from fronds of *Laminaria digitata*. The histogram columns correspond to the numbered circles on the frond. (From Stebbing, 1972.)

population on the Northumberland coast Daly (1978) found that individuals brooding embryos were apparent as early as April, and that a proportion of the population was still brooding in October; juveniles commenced settlement in June and small numbers were still settling in October. Mature individuals produced a succession of broods, at increasing frequency, through the whole of the summer.

Experiments with the larvae of epiphytic bryozoans have also shown that each species will choose to settle on the alga on which it normally occurs (table 2). Choice experiments using different algal species and surfaces conditioned with algal extracts, and more ambitious field experiments involving the monitoring of natural settlements of larvae on transplanted algae, have been conducted with numerous sessile epiphytes (see review in Hayward, 1980) and there is potential for much more work of this kind. It seems certain that chemical recognition of the substratum by the prospecting larva is an important stimulus to settlement, although so far the chemical substances involved have not been identified. This topic is of particular interest in considering the distribution of the epiphytes on the seaweed. Examination of a *Fucus serratus* plant will show that particular epiphytes appear to be concentrated on certain regions of the frond. Part of this effect, obviously, will be related to the growth of the plant, but experiments suggest that the larvae of some species settle at highest densities on defined areas of the plant. Ryland (1959) observed that *Alcyonidium* larvae settled most frequently on pieces of *F. serratus* tissue cut from just below the tips of vegetatively growing fronds, and avoided both basal tissue and the growing tips. An analysis of both naturally occurring and experimentally induced settlements of *A. hirsutum* and *A. gelatinosum* (Hayward & Harvey, 1974) confirmed this (fig. 25). The *Fucus* plant loses foliage basally, fruiting tips are shed annually, and vegetatively growing tips have a mucus-covered surface; thus, it seems that *Alcyonidium* larvae settling on the middle regions of the frond are maximising their chances for survival and growth. Stebbing (1972) conducted some elegant experiments with larvae of the tufted bryozoan *Scrupocellaria reptans* and two species of *Spirorbis* which occurred on fronds of *Laminaria digitata* at a sheltered locality on the west coast of Ireland. The larvae were offered a series of discs, 7.5 centimetres in diameter, cut from a *Laminaria* frond at successively greater distances from the stipe. The results were striking (fig. 26), showing that the larvae settled overwhelmingly on those discs cut from the basal regions of the frond. The *Laminaria* frond loses foliage from the tip, and is shed completely each spring, and grows by elongation of a basal region close to the stipe, the intercalary meristem. So, it would follow that epiphytes settled on or close to the intercalary meristem have the highest probability of survival.

The mechanisms underlying these behavioural patterns are still not clearly understood. However, they emphasise an interesting and ecologically significant fact: that the surface of an algal frond represents an age gradient. Whichever quality of the surface the larva is responding to, it is likely that the gradients of attractiveness revealed by settlement experiments reflect age-dependent gradients in the nature or concentration of settlement-inducing factors. Larval settlement experiments offer great scope to the experimental ecologist (see chapter 8) and it is desirable that the studies reported here be repeated with different populations and different species, of both algae and epiphytes, from a wider range of localities and ecological conditions. The interpretation of results is not easy and it should be remembered that marine larvae may be sensitive to, or sometimes quite oblivious to, a wide range of environmental stimuli. However, larval ecology is just one approach to the study of the distribution of sessile epiphytes and interesting data may be obtained also from the study of adult populations.

Fucus serratus is a perennial plant achieving a lifespan of more than 4 years in optimal habitats (Knight & Parke, 1950). Vegetative growth by apical elongation commences in the spring and accelerates through the summer, declining again in the autumn. Growth rates may be substantial, with vegetative fronds increasing by 3 centimetres monthly during the summer. Fruiting tips begin to appear in early spring and gametes are shed in early summer. During the late autumn and winter fronds with fruiting tips are shed and the plant loses foliage basally, the degree of loss often depending on winter storms. The density, age structure and individual plant morphology of a *Fucus serratus* population depend very much on environmental conditions. In optimum habitats large, richly branched and bushy plants will achieve maximum coverage of the middle and lower shore. Markedly suboptimal conditions result in sparse populations of short, little-branched and rather spindly plants, with consequent effects on the epiphyte populations. Algal substrata are unique in that they are self-renewing. Algal epifaunas differ from other sessile communities, such as those on intertidal rock surfaces, in that the provision of new habitat space, which is an important limiting resource for sessile communities, is not dependent upon disturbance, competition, predation or chance events (see Barnes & Hughes, 1982), but results from the growth of the alga (Seed & O'Connor, 1981).

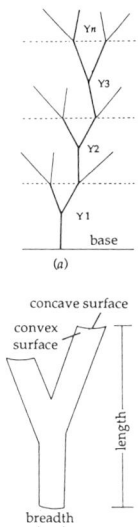

Fig. 27. (a) Division of a *Fucus serratus* frond into successive Y-segments. (b) Parameters of each segment.

The survival and structure of the epiphyte community are directly related to the structure of the algal community. A number of ecologists have investigated the composition of algal epifaunas in relation to the effects of varying environmental conditions on the algal population. For example, O'Connor and others, in a long series of papers (see Seed & O'Connor, 1981, for review), have studied the *Fucus serratus* community at localities in north and south Wales, and in Northern Ireland, employing simple standardised techniques for the collection of their data. Each plant studied was weighed whilst damp and the length of its longest frond measured. It was demonstrated that the ratio of plant weight to length gave a good estimate of bushiness, with high ratios indicating bushy plants and low ratios characterising elongate, spindly plants. The most heavily colonised plants from each locality were selected, and from each the longest frond was removed and cut into successive Y-shaped pieces (fig. 27) for detailed study. The area (A) of each Y-segment could be approximated by $A = 2BL$, where B was the greatest breadth and L the length. The abundance of the epifauna was recorded as numbers per Y-segment (in the case of solitary organisms) or percentage cover (in the case of bryozoans).

Table 6. *The comparative role of various environmental factors in influencing species occurrence on* Fucus serratus *in Strangford Lough, N. Ireland*

The number of signs indicates the relative favourability (+) or unfavourability (-) of factors. Zero indicates no apparent effect; brackets that the effect is slight.

	Sponges		Ascidians		Bryozoans				Tubeworm
	Grantia compressa	*Sycon ciliatum*	*Didemnum maculosum*	*Polyclinum aurantium*	*Flustrellidra hispida*	*Alcyonidium hirsutum*	*Electra pilosa*	*Membranipora membranacea*	*Spirorbis spirorbis*
Environmental characteristics									
High turbulence	++	+	++	-	++	+	(+)	--	-
Strong current	++	++	++	-	++	(+)	(-)	(+)	--
High silt loads	--	--	--	+	--	-	(-)	++	++
Plant characteristics									
Population density	++	++	++	+	++	(+)	+	(-)	+
Plant size	0	0	++	+	(+)	0	0	+[a]	(-)
Branching ratio	0	0	0	0	(-)	(+)	(+)	(+)[b]	(+)
Within-plant features									
Concave surface	++	(+)	-	-[c]	0	(+)	0	0	(+)
Basal position	++	++	++	+[d]	++	0	(+)	0	-
Y segment area	0	0	0	0	+	(+)	+	0	(+)

From Seed & O'Conner (1981).
[a] Size of plants utilised varies with competition intensity.
[b] Preference reverses in areas poorly colonised by other Bryozoa.
[c] On long plants only.
[d] Preference weak or negligible on concave surface.

Turbulence at each site (a measure embracing both current flow and abrasion effects) was estimated by reference to the percentage weight loss of gypsum spheres (technique p. 97), attached to three different levels of a plant on the shore and collected after one tidal cycle. Finally, whole plants bagged on the shore were removed to the laboratory and washed in known quantities of filtered seawater; using a Coulter counter, silt cover could be estimated from counts of particles (3-20 micrometres in diameter) in 0.5 millilitre subsamples from the washings. Using these data the investigators were able to examine the occurrence and within-plant distribution of the epiphytes in relation to algal population characteristics, individual plant morphology and environmental variables (table 6). The results suggested that each species achieved a degree of ecological isolation, occurring most abundantly at particular combinations of environmental, plant population and plant morphology characteristics. This type of segregation by habitat was considered to be important for the avoidance of spatial competition. That such competition does occur in the 25*F. serratus* community was shown by Stebbing (1973), who found that the bryozoan *Flustrellidra hispida*, which has relatively large, thick zooids, regularly overgrew and smothered the tubeworm *Spirorbis spirorbis*.

The life cycle of one *F. serratus* epiphyte, the bryozoan *Alcyonidium hirsutum*, was studied over a 3 year period at a locality on the Pembrokeshire coast by Hayward (1973). Larvae were released in February, settling higher on the frond than the parent colonies, and the young colonies grew rapidly through the early summer to form an interlocking mosaic, at which point further growth ceased (fig. 28). Growth of the alga continued through the late summer and early autumn ensuring new frond surface for the next generation of bryozoans. On most shores *Flustrellidra hispida* is a summer breeder, releasing its larvae in late May or June. Where *A. hirsutum* and *F. hispida* occur together they thus have partial ecological isolation through the timing of reproduction. On a single plant the colonies of the two species are therefore usually concentrated in distinct bands, with only narrow areas of overlap. The longevity of bryozoan epiphytes is, however, quite unknown and if it differs between species then opportunities may exist for one species to expand into new space resulting from the mortality of colonies of other species. Similarly, as particular features of the habitat (table 6) become unsuitable for one species then another may be enabled to expand its within-plant distribution. Further understanding of the dynamics of the *Fucus serratus* epifauna now requires long-term study of the growth and reproduction of the epiphytes in relation

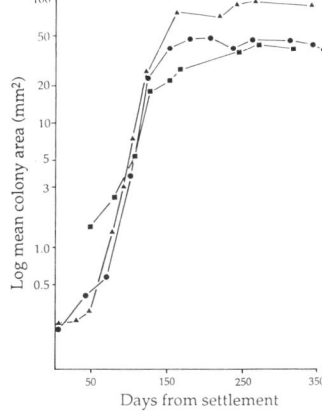

Fig. 28. Growth of *Alcyonidium hirsutum* in three successive years.

to the ecology and biology of the algal population.

The consequences for the alga of an increasing burden of sessile animals are unclear. It is probable that the fronds of heavily encrusted plants do not float so readily when covered by the tide, which would place them at a disadvantage with perhaps vigorously growing neighbours. The reduction in surface area available for photosynthesis might also be a problem, although as populations of many species of filter feeders tend to be located in the central regions of the plant this effect may be of less importance. As the epiphyte load increases large plants are likely to become increasingly unstable and particularly liable to storm damage during the winter. This aspect of seaweed ecology has scarcely been touched upon and is in need of good basic research. However, there is evidence of counteraction on the part of some algal species which perhaps serves to lessen the impact of colonisation by sessile organisms. *Himanthalia elongata* has a perennial, button-like stipe (fig. 29) which is usually firmly attached to the substratum. The top surface of the stipe constitutes the plant's meristem, and annually it grows a slender, whip-like frond. The under-surface of the *Himanthalia* button is often heavily colonised by sessile animals, including a number of bryozoans and the tubeworm *Spirorbis inornatus*. Field observations suggest that species diversity and density are highest on the larger buttons. The top surface of the button is invariably free of epiphytes. This is perhaps partly attributable to environmental effects; the alga is most abundant on exposed shores, and clearly the undersurface of the stipe offers the greatest degree of protection for the epiphytes. Al-Ogily & Knight-Jones (1977) conducted a series of choice experiments with *S. inornatus* larvae and found that they settled overwhelmingly on tissue from the basal surface of *Himanthalia* buttons in preference to distal meristem tissue. Pieces of algal tissue were then incubated on agar plate cultures of *Staphylococcus* bacteria and it was found that the meristem tissue, but not that from the base of the button, possessed an antibacterial property which resulted in zones of bacterial suppression around the cut pieces of alga. *Himanthalia*, then, is able to discourage the development of micro-organisms on the vulnerable top surface of its stipe. Whether its surface is thereby distasteful to prospecting *Spirorbis* larvae, or whether it simply inhibits the growth of the microbial films necessary to induce larval settlement, remains to be discovered.

A number of common intertidal algae are known to exhibit antibacterial activity on the surface of their fronds, and the highest levels of activity seem to be associated with the apical meristems of the plants (Hornsey & Hide, 1976).

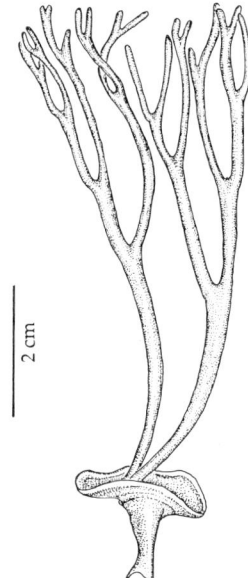

Fig. 29. *Himanthalia elongata* (L.) S.F. Gray: a young plant, showing basal holdfast button with apical fronds.

The result is an antifouling property, particularly marked on the growing tips of the plants, which tends to prevent the development of microbial films and, directly or indirectly, deters the settlement of invertebrate larvae. However, there is one curious anomaly in these results: fronds of *Laminaria digitata* and *L. saccharina* both showed surface antibacterial activity which increased along an age gradient from the basal region to the tip of the frond (Hornsey & Hide, 1976). This result correlates satisfactorily with Stebbing's work (fig. 26) and begins perhaps to explain the basis of the gradient of attractiveness to which many invertebrate larvae seem sensitive, and which is apparent on the fronds of many algal species. However, it is odd that in *Laminaria* the region of least antibacterial activity should be the basal intercalary meristem. It seems that in *Laminaria* surface fouling of the meristem is less significant than the reduction of photosynthesising frond area, whereas in fucoids the apical meristems are particularly vulnerable to damage through epifaunal settlement. Some seaweeds, including *Sargassum* (one species of which, *S. muticum* (Yendo) Fensholt, is now established at several localities on the south coast of England) and the more familiar *Fucus vesiculosus*, are known to exude mucus rich in tannins. In addition to possessing antibacterial properties, these tannins were found to affect the growth of mussels and barnacles when concentrated in tide pools (Conover & Sieburth, 1966). Algal responses to their epiphytic burdens are still scarcely researched and there is considerable scope, in both field work and laboratory investigation, for original contributions to this subject.

5 The fauna of kelp holdfasts

holdfast: the basal portion of an alga which fixes it to the substratum

haptera (singular hapteron): the rootlike structures which form the holdfast of the kelp plant

a polychaete worm

an isopod

an amphipod

a bryozoan

The kelp *Laminaria hyperborea* is the largest and ecologically most important of the British brown algae. It occurs subtidally off all rocky coasts, in dense monospecific forests, or intermingled with *L. digitata* and *L. saccharina*. Kelps have been widely studied and there is a considerable body of published data on their biology and ecology (see Kain, 1971). They are long-lived perennial plants and maintain their firm attachment to the rocky substratum by a substantial holdfast, which increases in size yearly as the plant grows. Kelp plants may be aged fairly accurately (fig. 13, p. 9); size and weight of the holdfast are easily measured and examined for correlation with the age of each sample. These data provide a basis for the comparison of kelp populations from different localities or different depths offshore, and also suggest a framework for the comparison of holdfast faunas.

As a *Laminaria* plant grows, its holdfast increases in size by the addition of successive concentric whorls of clasping rootlets, or haptera, which enclose increasing volumes of space. These spaces within the holdfast become occupied by a variety of animals. The surface of the haptera is usually colonised by encrusting, filter-feeding animals and the total number of species occurring in a good-sized holdfast is often extraordinarily high. Up to 50 species may be commonly expected. These include motile species, such as the polychaete worms* *Nereis* and *Harmothoe*, small crabs such as *Pilumnus hirtellus* (L.), *Pisidia longicornis* (L.) and *Carcinus maenas* (L.), isopods*, small snails, and even fish. Many crevice-dwelling invertebrates will live in holdfasts, including the isopod *Dynamene*, the lamellibranch *Kellia suborbicularis*, and sea cucumbers *Cucumaria*. Sedentary filter feeders or suspension feeders, such as the polychaetes *Sabellaria* and *Branchiomma*, mussels *Mytilus edulis* and *Musculus*, brittle stars *Ophiothrix fragilis*, and numerous tube-building amphipods*, may completely fill the spaces between the haptera. The community may be species rich, or may be dominated by one or a few species. Encrusting organisms include sponges and coelenterates, the barnacle *Verruca stroemia*, and many bryozoans*. There have been numerous surveys of kelp holdfasts and although the number of invertebrate species recorded from this habitat is impressive, none is known to be limited to it (with the exception of the Blue-rayed Limpet *Patina pellucida*). Thus, by the definitions arrived at in preceding chapters, no holdfast-inhabiting species may be properly termed epiphytic. Yet, the holdfast fauna constitutes a good sample

5 The fauna of kelp holdfasts

epibenthos: bottom-living marine animals, living on the surface of the sea floor, or on other surfaces such as rocks

of the epibenthos of a particular, narrow, marine habitat zone, and for the purposes of practical marine ecology the kelp holdfast is an ideal sampling unit. In view of the ubiquity of kelp beds and the richness of the holdfast fauna, it is natural to consider it together with other seaweed faunas.

Although the majority of any kelp population is subtidal, holdfasts may be collected from the lower shore on good spring tides. The stipe should be cut with a sharp knife about 20 centimetres above the top of the highest haptera (leaving enough stipe to permit accurate ageing of the plant), and a large plastic bag drawn over the holdfast, which is then cut from its substratum and eased into the bag, carefully, with all its contained fauna. Useful surveys may be conducted using shore-collected material, but, ideally, sample holdfasts should be collected at standard depth intervals by SCUBA divers, using the same 'cut and bag' technique outlined above. Jones (1971) studied holdfast faunas on the northeast coast of England and developed a method for estimating the volume of space within a holdfast available to the fauna. Observing that the holdfast approximated to a flat-topped cone with an elliptical base, he showed that the volume of the solid shape could be calculated as:

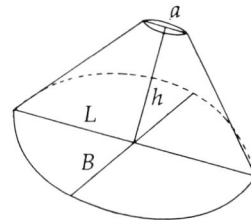

Fig. 30. Diagram of measured dimensions of a kelp holdfast. (After Jones, 1971.)

$$\frac{1}{12} \pi h \frac{B}{L} (L^2 + La + a^2) \text{ (measured in centimetres)} \quad (1)$$

where h is the height of the holdfast, B the breadth of the holdfast base, L the length of the holdfast base, and a the stipe diameter immediately above the holdfast (fig. 30). When the volume of the solid shape had been calculated, the holdfast was cut into pieces to remove all of the fauna, and the pieces weighed while wet. The volume of the holdfast tissue (in cubic centimetres) could be calculated as:

$$\text{wet weight of tissue (in grams)} \times 1.30, \quad (2)$$

where 1.30 is a constant representing the specific gravity of *L. hyperborea* tissue. Then, subtracting the product of equation (2) from that of equation (1) gives the volume of space, or 'habitat volume' (in cubic centimetres) within the holdfast.

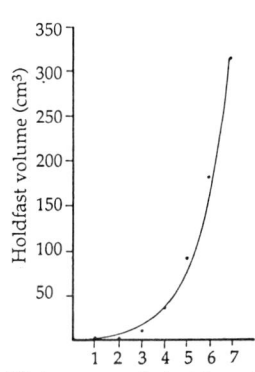

Fig. 31. Relationship between space within the holdfast in *Laminaria hyperborea* and age of the plant. (After Jones, 1971.)

In the population studied by Jones (1971) available space within the holdfast increased with the age of the plant (fig. 31), and the number of species occupying the habitat similarly increased. It was suggested that the fauna was recruited by larval settlement and migration from local epibenthos, and that the community developed in a successional manner, similar to that of plant communities.

trophic: pertaining to feeding

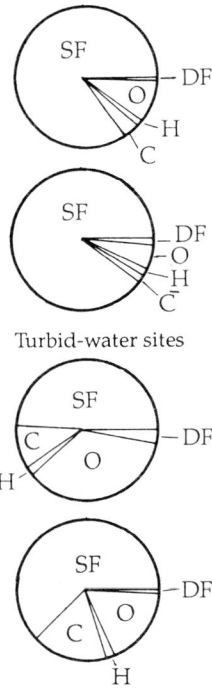

Fig. 32. Proportions of invertebrates in different feeding categories in holdfast faunas at turbid- and clear-water localities in southwest Ireland. SF, suspension feeders; DF, deposit feeders; O, omnivores; C, carnivores; H, herbivores. (From Edwards, 1980.)

Thus, the earliest colonisers, of 1-year old plants, were encrusting, filter-feeding bryozoans. In the next year these were succeeded by a number of suspension-feeding organisms, by omnivores, such as detritus-feeding amphipods, and finally by carnivorous species. As the holdfast increased in size with time its fauna became increasingly diverse, with its species showing many different kinds of feeding, or trophic specialisation. As the community approached climax some species, typical of early successional levels, declined or disappeared, while others, appearing at different stages in the development of the community, would persist. A major objective of Jones' (1971, 1973) work was to discover how both the diversity of the *Laminaria* holdfast community, and its development, were affected by coastal pollution. Results suggested that in heavily polluted areas species-rich, trophically complex communities were replaced by unstable, species-poor faunas dominated by suspension feeders. Subsequently, Moore (1974, 1978) showed that the turbidity of coastal waters may have a profound effect on the composition of holdfast faunas. On the Northumbrian coast holdfast faunas decrease in richness towards the silt-laden outflow of the River Tyne. As a pollution indicator, therefore, the utility of the holdfast fauna has still to be proven. However, Jones' (1971, 1973) work on the development and succession of kelp holdfast faunas is particularly interesting and has yet to be repeated elsewhere in such detail.

Moore (1974, 1978), also working along the northeast coast of England, paid particular attention to the amphipod populations of holdfast faunas. Major differences between localities, in terms of species diversity and the population density of each species, were related to silt loads (seawater turbidity), and turbid-water species groups could be distinguished from clear-water groups. However, in a larger survey of holdfasts from 11 points on the west and south coasts of Wales, Lundy Island and south Cornwall the same correlation was not evident. Instead, faunal differences were related to latitude, with distinct northern and southern species recognisable. Potential correlation with other factors was undoubtedly masked, in part, by the greater variety of habitat on the west coast, compared with the northeast coast. Two species of amphipod, *Lembos websteri* and *Corophium bonellii*, had been found on the Northumbrian coast to be highly susceptible to turbidity. In the west coast samples the same held true only for the former, while *C. bonellii* showed no significant difference in population density between clear-water and turbid-water sites. Studies on the feeding biology of *Lembos* showed that it collected some food by filter feeding, but also captured particles with its antennae. Population numbers were depressed by increased levels of clay-sized particles in its environment, which cut out light and perhaps prevented efficient feeding. *Corophium* was largely a filter feeder, and presumably able to cope with fine particles, and its feeding success was perhaps related to the nutritive value, and hence composition, of the silt load.

Water turbidity was found to be an important factor in holdfast faunal diversity on the coast of southwest Ireland (Edwards, 1980). In this study the data for each of 30 holdfasts collected at each sampling point were pooled to give an average of 1.5 litres of habitat space for each locality. Species diversity was found to be significantly reduced at the two most turbid sites, where the holdfast assemblage was dominated by a single species, the tubeworm* *Pomatoceros triqueter*. Suspension-feeding organisms predominated at the turbid sites, while trophically more varied communities were found at clear-water sites (fig. 32). Interestingly, in common with certain other low-diversity, suspension-feeder-dominated habitats, it was found that although species diversity remained at the same low level between two sampling dates 7 years apart, the nature of the community evidently varied and suggested marked instability: the dominant organisms in 1968 were saddle oysters, *Monia patelliformis* and *Heteranomia squamula*, with *P. triqueter* constituting only 7.9% of the holdfast fauna, while in 1975 the saddle oysters comprised only 6.2% of the fauna and *P. triqueter* was the dominant organism.

Laminaria holdfast faunas offer great scope for both the field ecologist and the experimentalist. The holdfast comprises a complex microhabitat which is sensitive both to broad environmental influences and to a range of microenvironmental factors. Clearly, the effect of any factor on the holdfast community may be a result of responses by the fauna as a whole, or of differential responses by each species. The role of biological interactions, such as competition, in the development of the community, and the importance of recruitment and persistence of species, have scarcely been examined. The holdfast fauna seems infinitely variable, yet the convenience of the habitat as a sampling unit of known age and predictable history makes it an intriguing object for ecological investigation.

a spirorbid tubeworm

suspension feeder: an animal which feeds on particles or organisms suspended in the water around it. It may do this by filter feeding, by catching or capturing items, or by 'vacuuming' close to the bottom

filter feeder: an animal which actively collects food particles or plankton from the surrounding water by creating a water current through or over its feeding apparatus

6 Predators

predator: broadly describes an animal which derives its nutriment from feeding on other animals

carnivore: literally, a meat eater. Generally used to describe the more spectacular predators

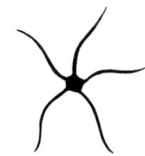

an echinoderm

a bryozoan

a sea spider

a hydroid

Predation is perhaps an important factor in the ecology of epiphytes, yet, although it offers a particularly absorbing field for investigation, there is very little published information on this topic. There are numerous records of field occurrences of known predators on presumed prey species. Some of these records are supported by direct observations of feeding, but there are few data on prey preference and feeding rates, practically no experimental observations on the interrelationships of predator and prey, and only a minimum of information on the life cycles and population ecology of known predators. Thus, this aspect of epiphyte ecology provides many opportunities for much-needed original work.

The very richness of some epiphyte communities - as exemplified by a well-encrusted *Fucus serratus* plant - probably makes them an attractive food source to a wide range of animals. Fish, echinoderms* and snails certainly graze such incrustations, and the small detritivorous and herbivorous snails are probably preyed upon by fish, crabs and carnivorous snails. However, in keeping with the emphasis of the preceding chapters, this section is concerned with those predators which feed mostly, or entirely, on the epiphytes discussed here. These include a number of specialised feeders, the life cycles of which may be closely attuned to those of their prey organisms. Most are rather small and individually inconspicuous, but their populations may be sufficiently high, and locally abundant, to make field recognition and the collection of material relatively easy.

The predators of the tiny snails so abundant in red algal turfs are largely unknown. The snails are probably eaten in large numbers by non-selective predators including fish (blennies), crabs and larger snails, and perhaps by the juveniles of certain carnivorous snails, such as the Dogwhelk *Nucella lapillus* (L.). However, although there is still little information available, there do not seem to be any specialised predators associated with these communities. Red algal turfs well encrusted with bryozoans*, and supporting good populations of snails, may also shelter large numbers of small sea spiders*, particularly *Achelia*. Others, including *Achelia, Anoplodactylus* and *Nymphon gracile*, are frequently found among the hydroid* *Dynamena pumila* and the bryozoan *Bowerbankia imbricata*, on *Fucus serratus*. The diet of most sea spiders is unknown; *Achelia longipes* has been observed grazing algal sporelings and *N. gracile* is known to feed on both *Dynamena* and *Bowerbankia*

(King, 1974). However, data on feeding, food preferences, population ecology and life cycles of these small predators are largely lacking and good observations on those associated with algal epifaunas are much needed.

Halacarid mites are often very common among seaweeds on sheltered shores. Among bushy *Fucus serratus* plants they may be found on colonies of *Alcyonidium hirsutum* and *Flustrellidra hispida*. On damaged areas of the bryozoan colonies they can be seen, together with harpacticoid copepods (fig. 40, p. 40), actually burrowing into dead or moribund zooids where they presumably feed on the bryozoan tissue. Whether, like the copepods, the mites are simply scavengers, or whether they are to be regarded as micropredators, remains to be resolved. The ecology of marine mites is a largely neglected field.

The most frequently studied group of predators, and consequently the one about which most is known, is the nudibranch sea slugs. Many of these distinctive, and often attractive, animals prey largely or exclusively on sessile filter-feeding organisms. Although the ecology of a majority of species still requires detailed investigation, many nudibranchs have been the subject of good ecological studies and there are several reviews on their biology and ecology, the most recent being that of Todd (1981).

nudibranch: the nudibranch sea slugs constitute the largest order of the Class Opisthobranchia (Fig. 47)

Table 7. *Preferred diets of some predatory sea slugs*

	Halichondria panicea	Hymeniacidon perleve	Myxilla incrustans	Alcyonidium hirsutum	Alcyonidium gelatinosum	Flustrellidra hispida	Electra pilosa	Membranipora membranacea	Dynamena pumila	Gonothyraea loveni	Halecium halecinum	Hydrallmania falcata	Kirchenpaueria pinnata	Obelia geniculata	Plumularia setacea	Botryllus schlosseri	Botrylloides leachi	Dendrodoa grossularia
Doridacea																		
Archidoris pseudoargus	Xs	X	X	-	-	-	-	-	-	-	-	-	-	-	-	-	-	-
Jorunna tomentosa	Xs	-	-	-	-	-	-	-	-	-	-	-	-	-	-	-	-	-
Acanthodoris pilosa	-	-	-	X	X	X	-	-	-	-	-	-	-	-	-	-	-	-
Adalaria proxima	-	-	-	-	X	X	Xs	X	-	-	-	-	-	-	-	-	-	-
Goniodoris nodosa	-	-	-	X	X	-	-	-	-	-	-	-	-	-	-	X	-	X
Goniodoris castanea	-	-	-	-	-	-	-	-	-	-	-	-	-	-	-	-	X	-
Onchidoris muricata	-	-	-	-	X	X	Xs	X	-	-	-	-	-	-	-	-	-	-
Polycera quadrilineata	-	-	-	-	-	-	X	Xs	-	-	-	-	-	-	-	-	-	-
Dendronotacea																		
Dendronotus frondosus	-	-	-	-	-	-	-	-	X	-	-	X	-	-	-	-	-	-
Doto coronata	-	-	-	-	-	-	-	-	X	-	-	X	-	X	-	-	-	-
Doto dunnei	-	-	-	-	-	-	-	-	-	-	-	-	Xs	-	-	-	-	-
Doto millbayana	-	-	-	-	-	-	-	-	-	-	-	-	Xs	-	-	-	-	-
Aeolidacea																		
Coryphella species	-	-	-	-	-	-	-	-	-	-	-	X	X	-	X	-	-	-
Eubranchus species	-	-	-	-	-	-	-	-	X	X	X	X	X	X	-	-	-	-

Modified from Todd (1981).   Xs = Specialist feeder.

Table 7, drawn from the more extensive tables of Todd (1981), lists some of the predatory nudibranchs known to feed on sessile epiphytes. An immediate difference is apparent between the food preferences of the Doridacea (rather solid, domed slugs which specialise principally on bryozoans) and the slender, often delicate Dendronotacea and Aeolidacea (which feed largely on hydroids). The majority of published data on food preferences in nudibranchs refer merely to the occurrence of the sea slug on its presumed prey species. For the more familiar species, including those listed in table 7, there are sufficient observations of actual feeding to be sure that the prey has been correctly recognised. However, it is often unclear to what extent a particular nudibranch may be a generalist or a specialist feeder; and if the latter, whether the preferred prey varies between populations. Certainly, the range of food species for some sea slugs appears to be very wide, while other slugs seem to be restricted to only a few prey species.

Adalaria proxima, studied by Thompson (1958), seems to be among the more specialised nudibranch predators. It is a cold-water species which on the west coast of Britain extends no further south than the Bristol Channel. It is often common on northwestern and eastern shores where it is found primarily on *Fucus serratus*. The Menai Strait population feeds principally on *Electra pilosa* (Thompson, 1958). A feeding *Adalaria* ruptures the frontal membrane of individual zooids of the bryozoan colony with short strokes of its radula and sucks out their contents with its muscular, pumping pharynx. *A. proxima* has an annual life cycle; juveniles appear on the shore in June, when the growth rate of their prey is at a peak, and feed continuously through the summer. Feeding declines in winter and the slugs begin to deposit their characteristic triple spirals of spawn on the seaweed in February. During April and May the adult slugs die, before the next generation settles from the plankton. Interestingly, although, in the absence of *Electra*, adults will feed on the other bryozoan epiphytes of *F. serratus* (*Flustrellidra hispida*, *Alcyonidium gelatinosum* and *Membranipora membranacea*), Thompson (1958) found that the larvae would delay settlement and metamorphosis until they came into contact with a live colony of *Electra pilosa*. When denied this stimulus they would prolong their planktonic life, and eventually cease to develop, and die. Todd (1981) noted that a number of other nudibranch species have equally specialised prey requirements and suggested that for these also, contact with the prey probably provided the essential stimulus to larval settlement and metamorphosis. However, the only other species in which similar experiments have been conducted is *Archidoris*

radula: a toothed chitinous ribbon forming part of the feeding apparatus of gastropod molluscs *pseudoargus* (pl. 5.4), which feeds on the sponge *Halichondria panicea*. In this case (Thompson, 1966) settling larvae displayed no response to the food species and the stimulus to settlement remains unknown.

Most species of nudibranch occurring in the intertidal zone either have annual life cycles or produce a number of short-lived generations in each year. The large dorids *Archidoris pseudoargus* and *Jorunna tomentosa* perhaps have biennial life cycles (Todd, 1981), although this was earlier disputed by Thompson (1958). In the case of *Archidoris* geographical variation in the timing of spawning confuses the picture and in some areas, such as the Clyde Sea and South Wales, spawning adults may be found in September. With the possible exception of *Archidoris*, all species have a single spawning period in late winter or early spring; juveniles appear in early summer and growth is continuous through summer to early autumn, when gonad development accelerates. All adults die shortly after spawning, not through reproductive exhaustion but from depletion of food reserves (Thompson, 1966).

There have been few population studies on intertidal nudibranchs, partly because of the problems of devising quantitative sampling techniques. Todd (1978) studied the distribution patterns of two populations of *Onchidoris muricata* on a shale shore at Robin Hood's Bay, Yorkshire. The slugs occurred on the undersurfaces of rocks where they fed on a number of encrusting bryozoan species.

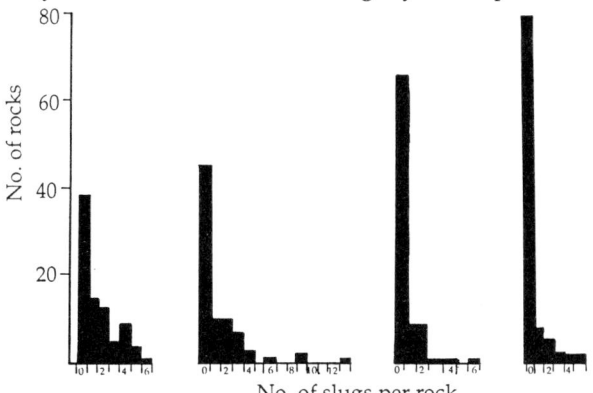

Fig. 33. Frequency of *Onchidoris muricata* on undersides of flat rocks, on four successive sampling dates. (After Todd, 1978.)

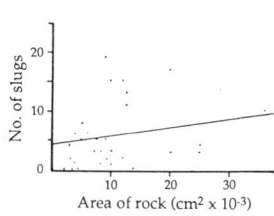

Fig. 34. Relationship between undersurface area of rock and number of *O. muricata* in a single sample. The regression slope ($y = a + bx$) was not significant. (After Todd, 1978.)

Each rock could be regarded as a standard sampling unit (fig. 33): size of the rock had no significant effect on the numbers of slugs present (fig. 34). *Onchidoris muricata* is frequently a common predator of *Alcyonidium hirsutum*, on *Fucus serratus*, and the population structure of the slug in this habitat would be worthwhile investigating, perhaps employing the *Fucus* plant as a sampling unit. Thompson

(1966) noted that the dorsal colour patterns of *Archidoris pseudoargus* are so variable that it was possible to identify individual animals on successive sampling occasions. The animals rarely move far when food supplies are plentiful and naturally isolated populations ought to be good subjects for long-term field monitoring.

Although field studies employing quantitative techniques will yield useful information on the population ecology of nudibranch predators, much more experimentation and laboratory observation are required to understand the finer details of their biology. Chadwick & Thorpe (1981) presented one of the few recent experimental investigations into the food preferences of three common species: *Polycera quadrilineata*, *Goniodoris nodosa* and *Onchidoris muricata*. Specimens of each nudibranch were offered a choice of presumed prey species in aquarium tanks and the time each individual spent on each food organism was recorded, with reference to the surface area of the food, and arranged to show a gradient of preference for each species of slug (table 8).

Table 8. *Prey preference in three nudibranch species*

Prey species are ranked in order of preference, giving percentage total observations per percentage area of substratum. One observation per 12 hours. 'Rock' means bryozoans on rocky substratum

	Polycera quadrilineata		*Goniodoris nodosa*		*Onchidoris muricata*	
1.	*Electra pilosa*	2.0	*Ascidiella scabra*	4.0	*Electra pilosa*	4.0
2.	*Alcyonidium gelatinosum*	2.0	*Botryllus schlosseri*	3.4	*Membranipora membranacea*	2.8
3.	*Membranipora membranacea*	1.5	Rock	2.0	*Celleporella hyalina*	2.0
4.	*Flustrellidra hispida*	1.1	*Alcyonidium gelatinosum*	1.4	Rock	1.5
5.	*Cellaria sinuosa*	1.0	*Flustrellidra hispida*	0.6	*Cellaria fistulosa*	0.8
6.	Rock	0.8	*Cellaria fistulosa*	0.2	*Cellaria sinuosa*	0.4
7.	*Cellaria fistulosa*	0.2	*Membranipora membranacea*	0.0	*Alcyonidium gelatinosum*	0.2
8.	*Scrupocellaria reptans*	0.0	*Electra pilosa*	0.0	*Scrupocellaria reptans*	0.0

From Chadwick & Thorpe (1981).

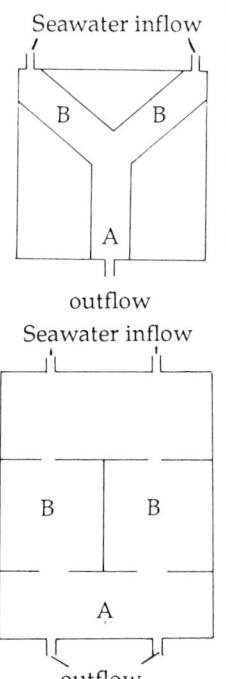

Fig. 35. 'Y' and 'T' maze apparatus used by Chadwick & Thorpe (1981). 'A' indicates the point at which nudibranchs were introduced; potential prey species or controls were sited at 'B'.

These preferences were further demonstrated in a series of experiments utilising T- and Y-mazes (fig. 35) in which the slugs were offered different food species in each chamber of the maze. *Onchidoris* showed a strong preference for the bryozoans *Electra pilosa* and *Membranipora membranacea*, *Goniodoris* preferred ascidians while *Polycera* was attracted to *Electra* and *Alcyonidium gelatinosum*. The narrow food preferences of predatory nudibranchs may help to prevent undue overlap in diet, and thus reduce competition between species. However, it is known that the preferred food species varies considerably between different populations of each nudibranch (table 7). The question of how food resources are partitioned in different populations and between varying suites of predators offers particularly good opportunities for research.

7 Identification

Introduction to the keys

The diversity of animals on the lower reaches of sheltered rocky shores is often astonishing and some experience is necessary before the beginner is able to distinguish epiphytic associations. A good field guide, such as Collins' *Pocket Guide to the Seashore* (Barrett & Yonge, 1958), is particularly useful, permitting rapid identification of the commoner organisms. The following keys are designed for the identification of species most of which are not included in popular field guides. They will allow identification of most sessile epiphytes, and also a high proportion of the sedentary and cryptic fauna of *Laminaria* holdfasts. The keys to sea spiders and sea slugs cover a majority of the presently known micropredators of sessile epiphytes together with the most important herbivores.

All animals collected should be examined in dishes of seawater, using a low-power stereomicroscope. A few species, once recognised, will be readily identifiable in the field, but the majority cannot be identified accurately without a microscope. It is often best to identify specimens while they are alive; indeed, many sea slugs are characterised by colour patterns which fade rapidly after death. However, bryozoans and hydroids are better appreciated when narcotised (Smaldon & Lee, 1979) and thus relaxed, and both amphipods and isopods usually have to be killed, by immersion in 70% alcohol, and viewed by substage illumination, before they can be identified.

The following notes, together with the Quick-Check Key, describe all of the major groups of animals associated with seaweeds and refer the user to a numbered series of keys. Not all animals encountered, particularly in holdfasts, are easily identifiable. For example, nematodes may occur commonly in holdfasts, as in every other aquatic and terrestrial habitat, but there is no complete monograph of intertidal species. Other groups, such as the harpacticoid copepods, require specialist knowledge and extensive libraries for correct identification. Such groups are not covered in these keys. However, all of the most frequent, and perhaps ecologically most significant, seaweed-associated species have been included. It should be stressed that non-epiphytic species will not key out; if any animal cannot be identified using these keys, or does not match the illustration of it, then it is not primarily an epiphyte and should be identified using the additional works referred to at the beginning of each key.

Begin your identification by using the Quick-Check Key (pp. 45-7). Check that you have arrived at the right group by reading the notes (pp. 39-44). The notes tell you which key to go to next.

In the keys, look first at the features at the beginning of each lead; these are the contrasts between the two alternative leads. Confirmatory characters are given afterwards.

Notes on major animal groups associated with seaweeds.

A. *Sponges*

a hydroid

Solitary, in the form of flat or cylindrical sacs with terminal oscula (fig. I.1) up to 10 millimetres long (pl. 1.1, 1.2); or colonial, forming irregular, rather structureless incrustations with numerous oscula. Often brightly coloured. **Key I**.

B. *Hydroids or sea-firs*

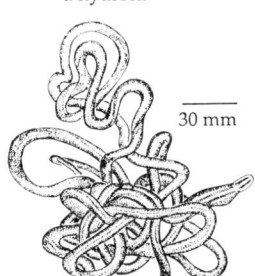

30 mm

Colonial coelenterates with erect, branching colonies; regular or irregular, feathery, tufted or diffuse. Colony with a continuous outer covering, or perisarc; hydranth (fig. II.3) immersed within the stem (pl. 1.3, 1.4) or stalked and projecting from it (pl. 1.5, 1.6), enclosed within a cup of perisarc, or free. On fucoid algae and *Laminaria*. **Key II**.

C. *Nemerteans or ribbonworms*

Fig. 36. *Lineus longissimus* (Gunnerus). (After Gibson, 1982.)

Soft unsegmented worms, usually with a mucus covering (fig. 36). Sometimes brightly coloured and often with distinctive eyespots. May be encountered in *Laminaria* holdfasts. A Linnean Society Synopsis (Gibson, 1982) describes all British species.

D. *Nematodes or roundworms*

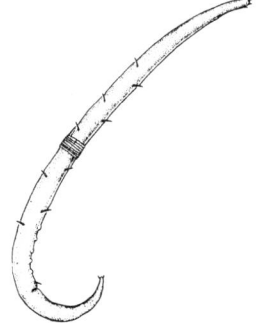

Fig. 37. Generalised marine nematode. (After Platt & Warwick, 1983.)

Unsegmented worms. Small (usually less than 1 centimetre), dull white, turgid, and typically stiffly coiled (fig. 37). Free-living and parasitic species may occur in holdfast communities. They are not easily identified, although size, presence/absence of setae, and buccal structure may permit recognition of different species in a sample. A Linnean Society Synopsis (Platt & Warwick, 1983) deals with one sub-order of free-living marine species, and describes techniques for collection and preservation.

E. Entoprocts

Colonies of tiny zooids (less than 2 millimetres) rising from branching, creeping stolons. Each zooid comprises a slender, cylindrical peduncle bearing a globular, tentaculate calyx. On fucoid algae, kelp holdfasts and stipes, and among bryozoan turfs. Two species are commonly encountered: *Barentsia gracilis* (fig. 38) with slender, nodulated peduncle; *Pedicellina cernua* (fig. 39) with stout, cylindrical, bristly peduncle.

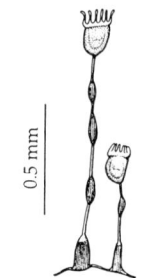

Fig. 38. *Barentsia gracilis* (Sars).

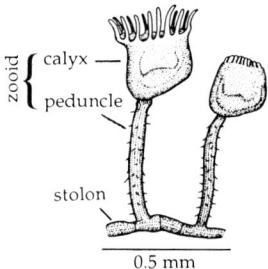

Fig. 39. *Pedicellina cernua* (Pallas).

F. Polychaetes or bristleworms

Segmented worms. Living in tubes of cemented sand and detritus, or secreting white, calcareous, sometimes spiralled tubes (fig. 8, p. 6), or free-living in crevices among holdfasts. Head with a crown of tentacles, or with various appendages, and sometimes powerful jaws. Body segments with few or many appendages; with or without dorsal scales, or setae (bristles) or cirri. On many different algae; sedentary on the frond, stipe and holdfast, or free-living in kelp holdfasts. **Key III.**

G. Copepods

Tiny crustaceans (less than 1 millimetre long), the body cylindrical or tapered, all segments very similar. Two pairs of antennae, the second pair long and obvious; other appendages not obvious from above; usually with elongate caudal rami. Harpacticoid copepods (fig. 40) are sometimes abundant on *Fucus serratus*, in sheltered locations, or in kelp holdfasts, where they feed on detritus. There is no easy means of identification.

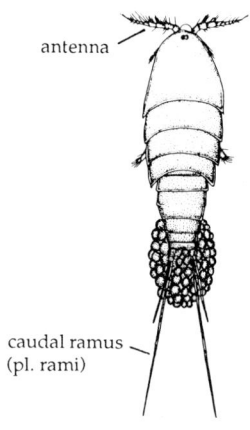

Fig. 40. *Harpacticus uniremis* Kroyer, a harpacticoid copepod. Female with egg mass.

H. Barnacles

Shelled, sessile crustaceans. Thoracic appendages, modified for filtering, are protruded when the animal is feeding under water. Shell symmetrical, with six plates and an operculum of four plates; or asymmetrical, with four unequal plates and an operculum of two plates. Two species occur not infrequently on kelp holdfasts: the conical, symmetrical *Balanus perforatus* (fig. 41), and the flattened, asymmetrical *Verruca stroemia* (fig. 42). All British intertidal barnacles may be identified using Southward (1976).

7 Identification

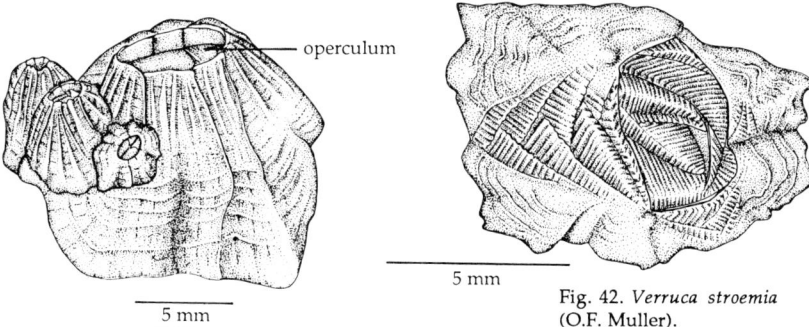

Fig. 41. *Balanus perforatus* Bruguière.

Fig. 42. *Verruca stroemia* (O.F. Muller).

I. Isopods

an isopod

Small (less than 2 centimetres long), flattened as seen from above; often elongate, but sometimes almost cylindrical (pl. 2.5-2.8). Two pairs of antennae, the second pair well developed; five to seven pairs of thoracic limbs, all more or less similar; five pairs of two-branched pleopods (fig. IV.1), one pair of two-branched or one-branched uropods, sometimes modified. Living in tufted or bushy seaweeds, particularly *Fucus* species, aggregated on detached drift weed, or in crevices among large holdfasts. **Key IV.**

J. Gammaridean amphipods (Sandhoppers and their allies)

Small crustaceans (less than 2 centimetres long) compressed from side to side. Two pairs of antennae; seven pairs of thoracic limbs, the first two pairs modified as gnathopods (fig. 43); three pairs of two-branched pleopods; three pairs of two-branched uropods. Free-living, or building tubes of sand, silt or detritus; on a wide range of algae, in red algal turfs, or in holdfasts, often occurring in dense populations. **Key V.**

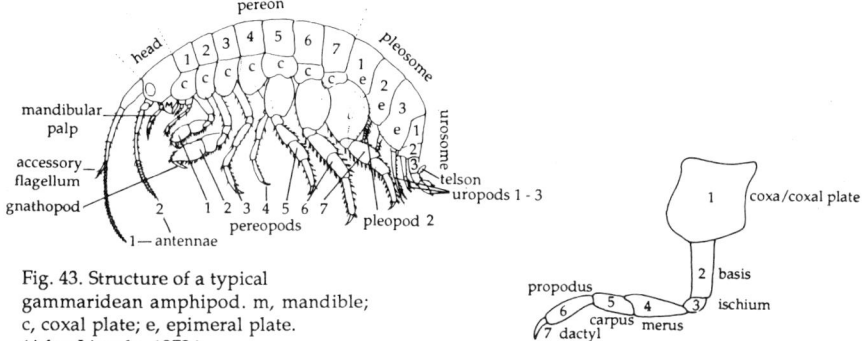

Fig. 43. Structure of a typical gammaridean amphipod. m, mandible; c, coxal plate; e, epimeral plate. (After Lincoln, 1979.)

K. *Caprellidean amphipods or skeleton shrimps*

Small crustaceans (less than 2 centimetres long), elongate, cylindrical (pl. 2.1-2.3). Body with seven segments, excluding head; two pairs of antennae; two pairs of gnathopods; three pairs of hooked legs behind (fig VI.1). Slow moving; attached to hydroids, bryozoans, or small algae. **Key VI.**

L. *Ostracods*

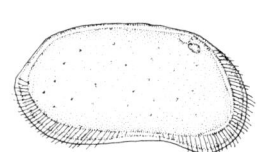

Fig. 44. An ostracod, *Cythere albomaculata* Baird. Male from right side.

Small crustaceans (less than 5 millimetres long), the body completely enclosed in a bivalved shell (fig. 44), limbs reduced and modified. Two pairs of antennae; one or two pairs of reduced legs are occasionally visible. Among bushy seaweeds, on detritus layers or other epiphytes, or in holdfasts. Ostracods are sometimes quite abundant in sheltered, detritus-rich habitats. There is no modern handbook on marine species.

M. *Pycnogonids or sea spiders*

Slow-moving, long-legged animals. Flattened, segmented body with four pairs of walking legs; head with a stout proboscis, and with or without paired palps and/or chelifores (fig. 45). One pair of egg-carrying appendages (ovigers) on first body segment in males. Up to 15 millimetres long. On bryozoans and hydroids, or among small seaweeds. **Key VII.**

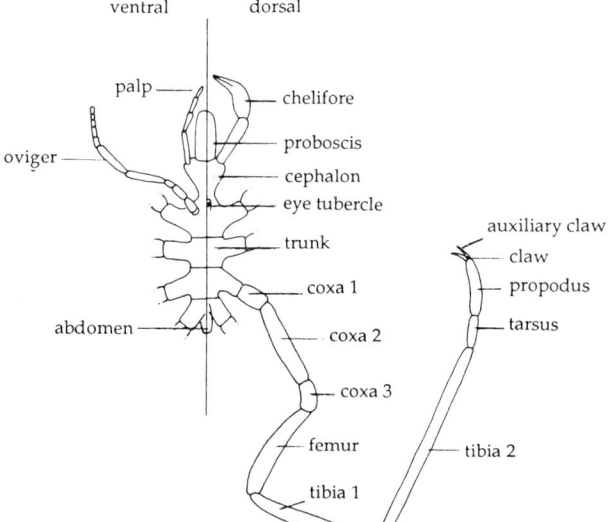

Fig. 45. Structure of a typical pycnogonid. (After King, 1974.)

N. Acari or (marine) mites

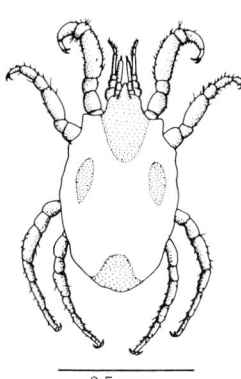

Fig. 46. A marine mite, *Halacarellus basteri* Johnston.

Most belong to a single family, the Halacaridae. Body oval or round, without visible segmentation, often brightly coloured. Four pairs of legs. Up to 5 millimetres long (fig. 46). Among small seaweeds, in holdfasts, on bryozoans, hydroids and other epiphytes. Both herbivorous and predatory species occur; identification is not easy and ecology is little studied. There is a French monograph (Andre, 1946) and a Linnean Society Synopsis on the British species (Green & Macquitty, 1987).

O. Chironomids (midges)

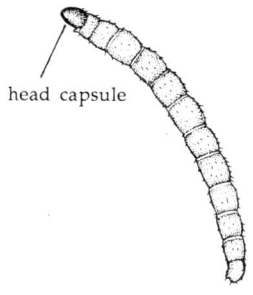

a larval midge, *Clunio*

Body cylindrical, segmented, up to 6 millimetres long, with sparse bristles; no obvious legs, but short appendages (parapods) at head and tail end; front end with distinctive head capsule. The larvae of most midges live in freshwaters, but those of a few genera (e.g. *Clunio*) live among middle and upper shore seaweeds and are sometimes very common. Bryce & Hobart (1972) and Cranston (1982) provide keys for identification which cover marine chironomid larvae.

P. Lamellibranchs or bivalves (mussels, cockles and their allies)

byssus: in some bivalves (e.g. mussels), a bundle of chitinous threads secreted by a gland on the foot, which attaches the animal to its substratum

Free-living or attached by byssus threads or cement. From less than 1 millimetre to greater than 20 millimetres long. In *Lichina* and small, bushy seaweeds, and in holdfasts. **Key VIII.**

a gastropod mollusc

Q. Prosobranchs, gastropod molluscs (snails, limpets and whelks)

Limpets: 2-15 millimetres long, on frond, stipe and holdfast of kelp. Snails: less than 1 millimetre to greater than 20 millimetres long, in *Lichina* and small red seaweeds, on bushy fucoids and in kelp holdfasts. **Key IX.**

R. Opisthobranchs or sea slugs

With reduced or internal shell, or rarely shelled, typically soft-bodied and lacking a shell. Body supported by a muscular foot; upper surface smooth or knobbly, with or without paired rhinophores at the front, circlet of branching gills at the back, or slender or branched cerata (fig. 47). On red and green seaweeds, on bryozoans, hydroids or other sessile epiphytes. **Key X.**

44   7 Identification

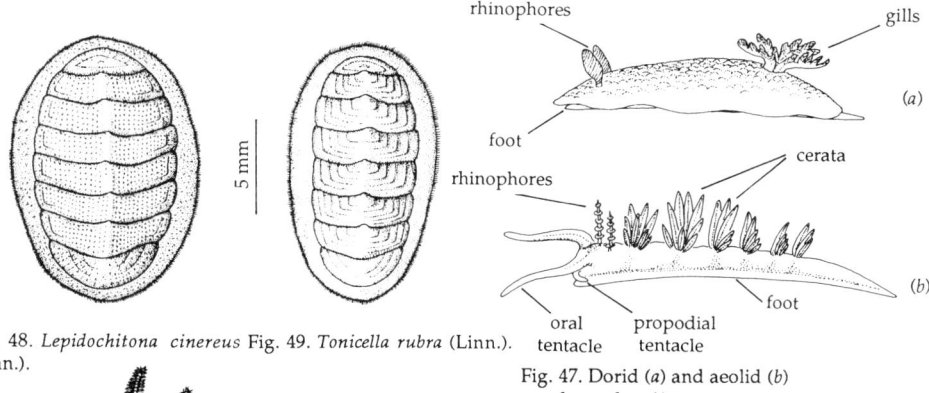

Fig. 48. *Lepidochitona cinereus* (Linn.).

Fig. 49. *Tonicella rubra* (Linn.).

Fig. 47. Dorid (*a*) and aeolid (*b*) sea slugs, showing morphological characteristics.

S. Chitons

Bilaterally symmetrical molluscs; with a grasping muscular foot and a shell of eight transverse plates, surrounded by a tough, granular, mantle border. Up to 20 millimetres long. Two common species may occur occasionally on kelp holdfasts: *Lepidochitona cinereus* (fig. 48), with rough-surfaced shell and finely granular mantle bearing cigar-shaped marginal spines; and *Tonicella rubra* (fig. 49), smooth-shelled, mantle covered with small, ovoid granules and with flattened marginal spines.

T. Bryozoans or moss animals

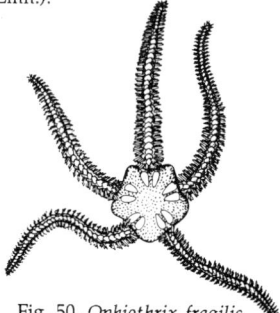

Fig. 50. *Ophiothrix fragilis* (Abildgaard).

Colonial animals with calcareous or non-calcareous, fleshy or gelatinous colonies; individual zooids visible with a x10 hand lens. Forming flat incrustations, or erect, bushy or spiralled tufts, or dense tangled mats. Common on *Fucus serratus*, *Laminaria* and numerous other seaweeds. **Key XI.**

U. Echinoderms

Brittle stars, particularly *Ophiothrix fragilis* (Abildgaard) (fig. 50), and sedentary, sausage-shaped sea cucumbers (fig. 51) may occur in larger holdfasts. All British species may be identified using Mortensen (1977).

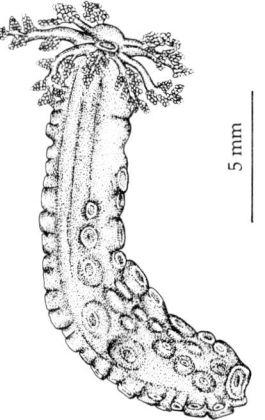

Fig. 51. A crevice-dwelling sea cucumber, *Cucumaria*.

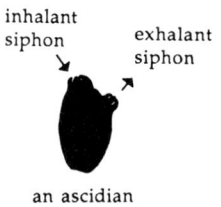

an ascidian or sea squirt

V. Ascidians or sea squirts

Colonial or solitary, brightly coloured incrustations with zooids in star-shaped groups or extended double chains (pl. 6.1, 6.3); or squat, leathery, wrinkled sacs with two siphons (openings); or soft, fleshy, round or stalked colonies of numerous zooids. On the fronds of fucoids and kelps, or attached to kelp holdfasts. **Key XII.**

Quick-Check Key
1. ANIMALS WITH LEGS
(a) Eight Legs

Phylum: Chelicerata
Class: Pycnogonida
   M. Sea spiders fig. 45
Up to 15 mm long. Body long and narrow, typically segmented; head with a cylindrical proboscis. **Legs thin, spider-like, usually much longer than body.**

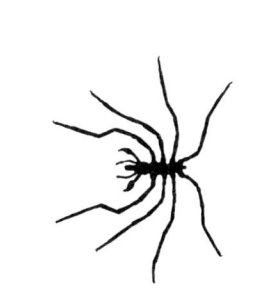

Phylum: Chelicerata
Class: Arachnida
Order: Acari
   N. Mites fig. 46
Up to 5 mm long. **Body round or oval, typically flattened, no obvious segmentation but with a few flat plates on upper surface. Legs not much longer than body.**

(b) More or fewer than eight legs

LEGS ALL ALIKE		LEGS NOT ALL ALIKE	

Phylum: Crustacea
Class: Copepoda
Order: Harpacticoida
   G. Copepods fig. 40
Body long and tapering, round in section; head with one pair of long and one pair of short antennae. Very small legs present on front end of body; hind end with slender forked 'tail'.
(Bristle worms may come out here if the projections along the body are mistaken for legs.)
	Phylum: Crustacea
	Class: Malacostraca
	Order: Isopoda
	   I. Isopods (pl. 2.5-2.8)
	Body flattened from top to bottom; with head, segmented middle region bearing legs, and distinct tail portion; first pair of antennae very short, second pair often very long.
		Phylum: Crustacea
		Class: Malacostraca
		Order: Amphipoda
		Suborder: Gammaridea
		   J. Sand-hoppers and their allies fig. 43
		Body flattened from side to side; with head, segmented middle region bearing legs, and distinct tail portion; both pairs of antennae clearly seen, of variable length. First two pairs of legs with claws or pincers; relative length of all legs very variable

			Phylum: Crustacea
			Class: Malacostraca
			Order: Amphipoda
			Suborder: Caprellidea
			   K. Skeleton shrimps (pl. 2.1-2.3)
			Body long and thin, consisting of head and 7 segments. Two pairs of pincers at front end, and 3 pairs of short legs at rear, often with shorter limbs between.

2. ANIMALS WITHOUT LEGS AND WITH SHELL
(a) Shell with two hinged valves

Phylum: Mollusca
Class: Bivalvia
P. Mussels, cockles and their allies

Molluscs up to > 20 mm long with body flattened from side to side and enclosed by two folds of tissue, the mantle, each secreting a hard shell; the two shell valves hinged dorsally. Head absent; gills well developed; a **triangular, muscular foot** characteristic.

Phylum: Crustacea
Class: Ostracoda
L. Ostracods fig. 44

Small (up to 5 mm long) bean-shaped shells; animal **swimming freely**; two small antennae projecting at the front, tips of two small legs at rear.

(b) Shell not with two hinged valves

Phylum: Mollusca
Class: Gastropoda
Sub Class: Prosobranchia
Q. Snails, limpets and whelks

Molluscs with an asymmetrical body enclosed in a (usually) coiled shell. Head well developed, with tentacles and eyes. Animal crawls on an elongate flat-soled foot.

Phylum: Mollusca
Class: Polyplacophora
S. Chitons figs. 48, 49

Molluscs with an oval body covered by a shell of 8 curved plates, bordered by a distinct mantle skirt. Lower surface with a powerful, muscular foot, holding the animal tightly on to the substratum.

Phylum: Crustacea
Class: Cirripedia
H. Barnacles figs. 41, 42

Highly modified **crustaceans** living attached to solid substrata, the body enclosed by a shell of 4-8 plates, closed by **2-4 hinged flaps**. Legs extended through aperture of shell **as a feeding fan**.

3. ANIMALS WITHOUT LEGS OR SHELL
(a) Worms

Phylum: Annelida
Class: Polychaeta
F. Bristle worms

Body divided into numerous segments, each bearing bristly appendages and often with various leaf-like or thread-like flaps. Head with tentacles, eyes and jaws, or with numerous thread-like tentacles. Free-living and mobile, or living in a tube of gravel or mud, or calcium carbonate.
(Chironomids may come out here, but have a head capsule or rigid casing to the head, and < 12 segments.)

Phylum: Nematoda
D. Roundworms fig. 37

Body unsegmented and rather stiff, without legs or bristles. Free-living or parasitic.

Phylum: Nemertea
C. Ribbon worms fig. 36

Body unsegmented, soft, thin, flat, sometimes very long, usually slimy. No obvious head, but eyes may be present at front.

Quick-Check Key

(b) Not worms

ANIMALS SINGLE	ANIMALS JOINED IN CLUMPS OR COLONIES	
Phylum: Mollusca **Class:** Gastropoda **Sub Class:** Opisthobranchia R. Sea slugs fig. 47 Molluscs with a more or less symmetrical body; shell small or internal, more usually absent. Head with eyes and tentacles. Upper side of body often with groups of gills and finger-like cerata; lower side with elongate crawling foot.	**Phylum:** Coelenterata **Class:** Hydrozoa B. Hydroids or seafirs Colonies of stalked zooids, loosely associated or forming ordered arrangements in complex branching growths. Colony typically with a continuous horny outer covering, the perisarc; zooids linked by common body cavity; bearing slender, non-ciliated tentacles. Reproductive stages commonly brooded in vase-shaped or bulbous receptacles larger than feeding zooids.	**Phylum:** Bryozoa T. Moss animals Colonies of encrusting sheets, diffuse or dense clumps, or more solid nodules, calcareous or gelatinous. Zooids with ciliated tentacles and an independent looped gut; colonies clearly formed of distinct box-like units, without a continuous outer covering.	**Phylum:** Entoprocta E.Entoprocts figs. 38, 39 Colonies of stalked zooids on a creeping stolon; each zooid with a swollen body, bearing short stumpy tentacles, on a slender, knobbly or bristly stalk. Not calcified; without a continuous horny outer covering.
	COLONIES WITH SMALL TENTACLE-BEARING ZOOIDS	
Phylum: Echinodermata **Class:** Stelleroidea U. Starfish and brittle stars fig. 50 Star-shaped animals with stiff bodies formed of calcareous plates and spines; with a central disc bearing the mouth on its lower surface, and 5 or more flexible arms.	**Phylum:** Chordata **Sub phylum:** Urochordata **Class:** Ascidiacea V. Sea squirts Living in clumps of oval or rounded bodies; or in flat sheets with the zooids in regular lines or stars. Each zooid with two openings for exhalant and inhalant water currents.	**Phylum:** Porifera A. Sponges In sheets, clumps or lobes, or as individual bottle or vase-shaped growths. Smooth, bristly, or distinctly spongy, but without other clear morphological features.	
Class: Holothurioidea U. Sea cucumbers fig. 51 Body cylindrical, stiff; with five longitudinal rows of tube feet; mouth at front end surrounded by radiating, branched tentacles.	WITHOUT TENTACLES	

I Sponges

A good account of the structure, biology, ecology and classification of sponges is provided by Bergquist (1978). The common fouling species are illustrated in colour by Sara (1974), but there is no modern handbook for the identification of British species.

1 Sponge in the form of a short, stiff cylinder or barrel; or a thin, flat sac; or tubular and branched, forming a dense clump; not forming a continuous encrusting sheet. Rarely exceeding 1 cm high; oscula always single, terminal, conspicuous (I.1). Off-white or yellowish 2
— Sponge in the form of a continuous encrusting sheet, with numerous oscula. Green, orange, yellow or red 4

2 Forming slender branching cylinders, often densely packed together. Common on *Fucus serratus*; sometimes on other large, lower shore, brown algae   *Leucosolenia* species (I.1)
— Not branching. Single, or clumped, but not forming a sward 3

3 Sponge vase-shaped or cylindrical, the osculum surrounded by a fringe of stiff bristles. On *Fucus serratus*, and other lower shore algae, in sheltered habitats, or among small red algae
Scypha ciliata (Fabricius) (pl. 1.1)
— Sponge flat, oval or irregular, with a smooth-rimmed osculum. Widespread and common, on many lower shore algae; often abundant among small red seaweeds
Scypha compressa (Fabricius) (pl. 1.2)

4 Smooth-surfaced; oscula regularly spaced, with distinct raised rims. Commonly green, sometimes yellow or orange. Widespread and common, on a range of substrata, on middle and lower reaches of rocky shores. Frequently densely intergrown with red algal turfs, and often developing extensive incrustations on kelp holdfasts   *Halichondria panicea* (Pallas) (I.2)
— Rough-surfaced; oscula densely and irregularly distributed. Commonly orange to red. Lower shore, predominantly on hard substrata, but not infrequent on kelp holdfasts
Hymeniacidon perleve (Montagu) (I.3)

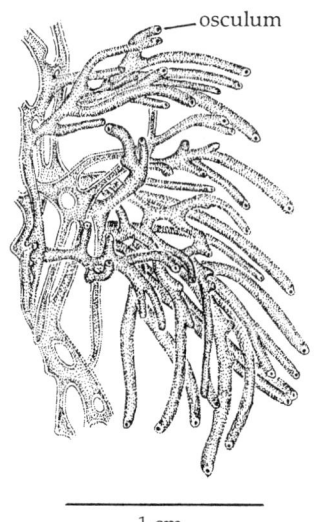

I.1   1 cm

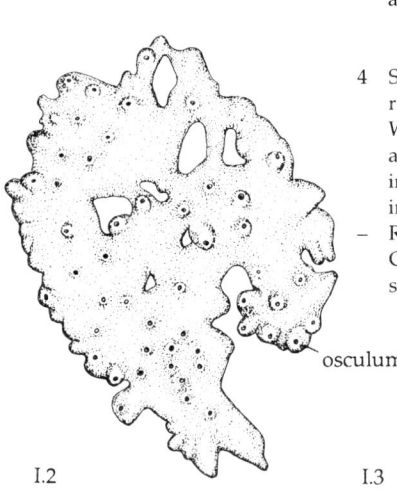

I.2   I.3   1 cm

PLATE 1

1. *Scypha ciliata* (Fabricius)
2. *Scypha compressa* (Fabricius)
3. *Dynamena pumila* (Linn.) part of colony
4. *Dynamena pumila* (Linn.) detail of hydrothecae
5. *Obelia geniculata* (Linn.) detail of hydrothecae
6. *Obelia geniculata* (Linn.) part of colony

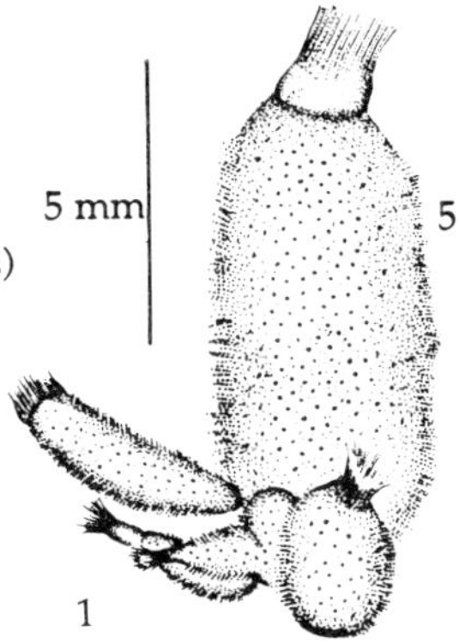

1

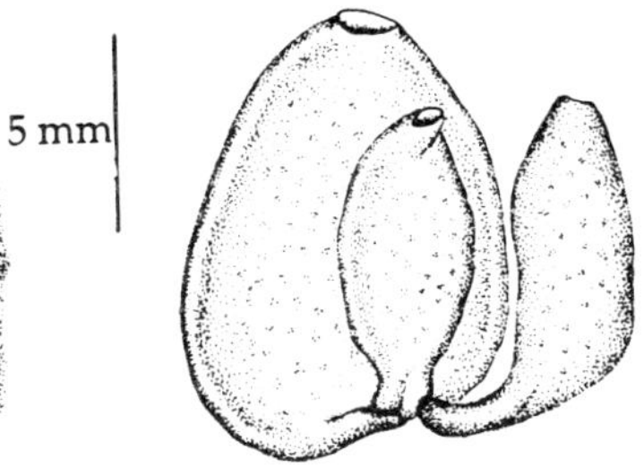

2

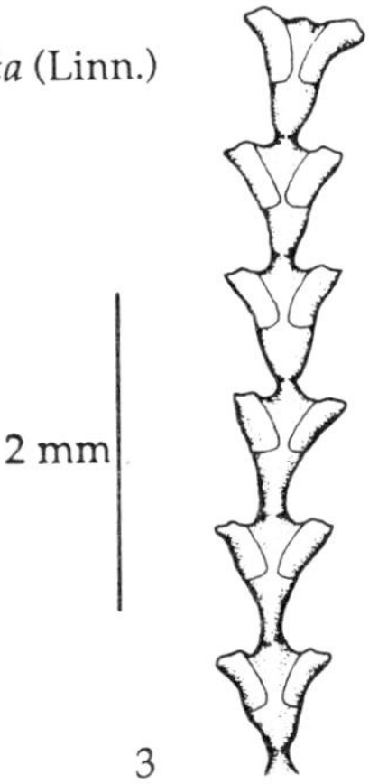

3

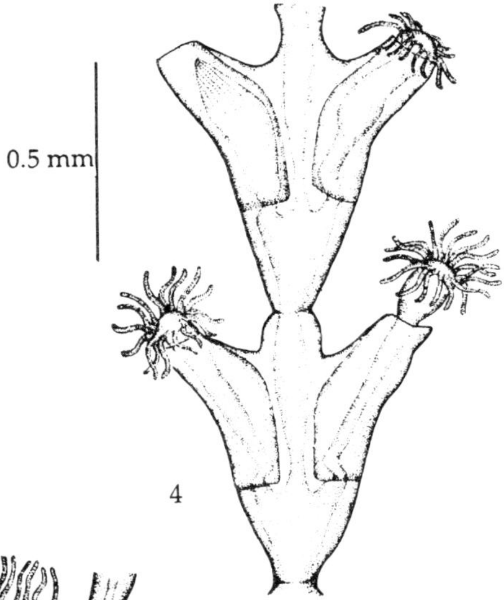

4

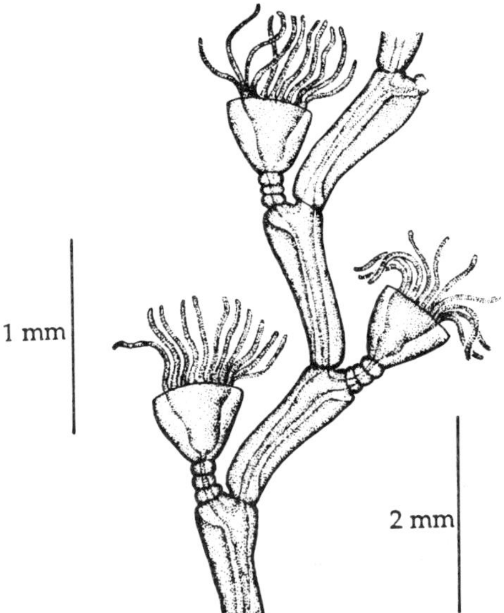

5

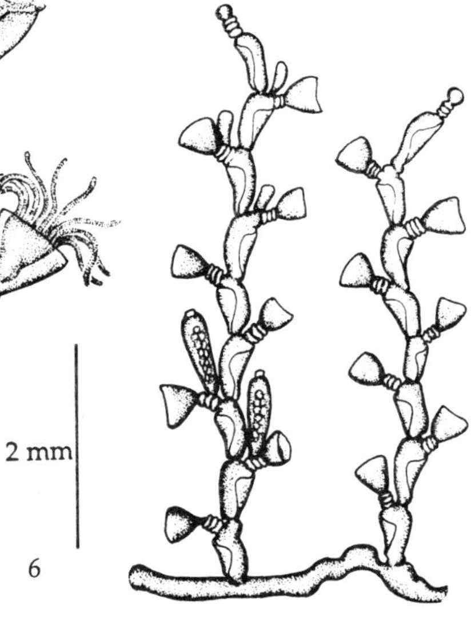

6

PLATE 2

1. *Caprella linearis* (Linn.) male
2. *Caprella linearis* (Linn.) female
3. *Pseudoprotella phasma* (Montagu) female
4. *Chaetogammarus marinus* (Leach)
5. *Munna kroyeri* Goodsir
6. *Idotea baltica* (Pallas)
7. *Jaera nordmanni* (Rathke) female
8. *Jaera nordmanni* (Rathke) male

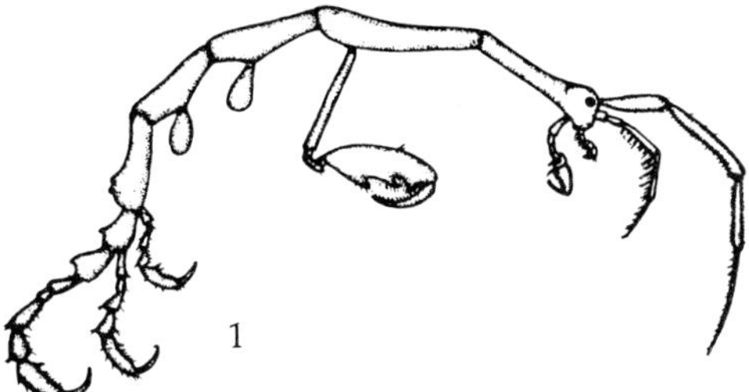

1

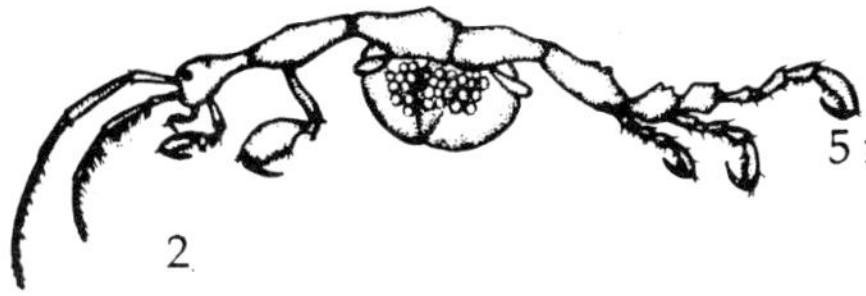

2

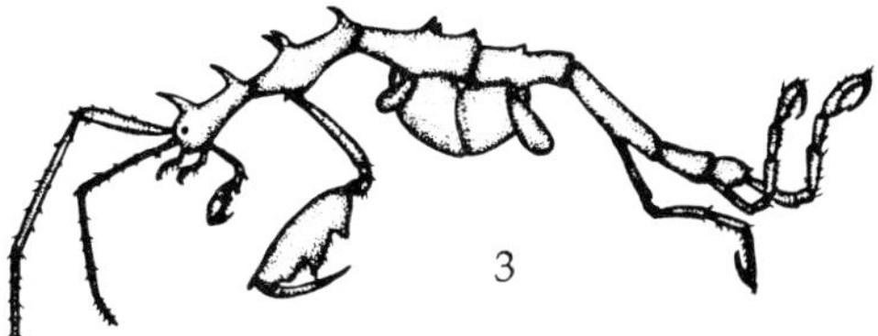

3

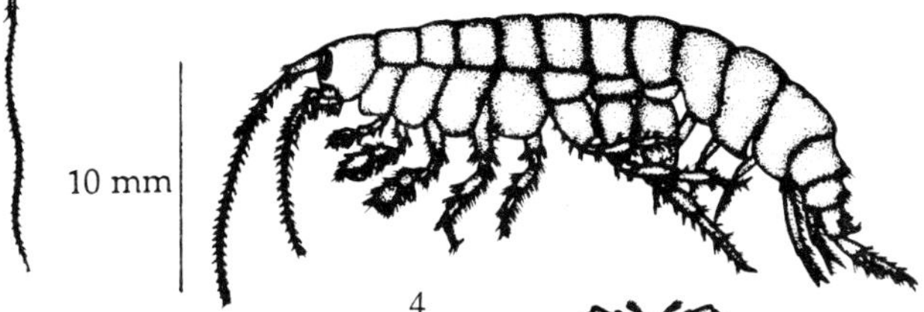

4

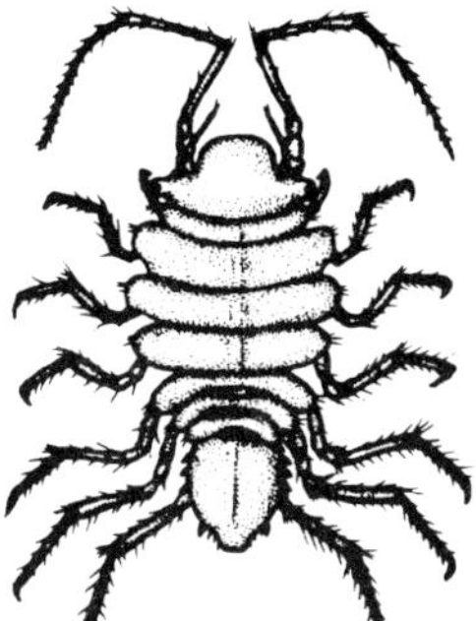

5

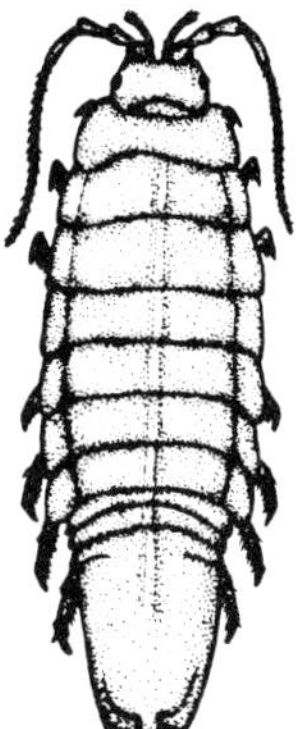

6

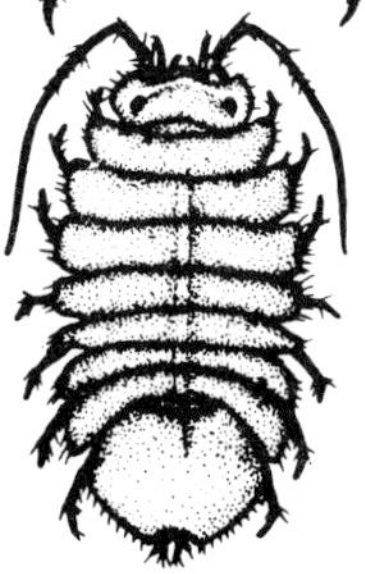

7

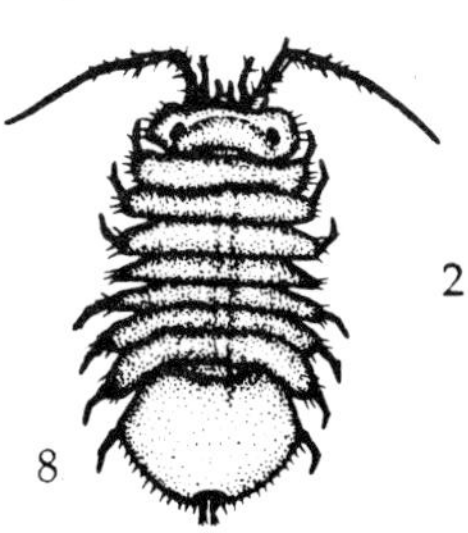

8

PLATE 3

Common brown seaweeds

1. *Laminaria digitata*
   Kelp

2. *Fucus vesiculosus*
   Bladder Wrack

3. *Ascophyllum nodosum*
   Knotted Wrack

4. *Fucus serratus*
   Toothed Wrack

Fig. 1 is one-eighth natural size.
Figs. 2-4 are one-third natural size.

PLATE 4

Common red seaweeds

1. *Lomentaria articulata*
2. *Corallina officinalis*
3. *Laurencia pinnatifida*
   Pepper Dulse
4. *Gigartina stellata*
5. *Chondrus crispus*
   Irish Moss
6. *Palmaria palmata*
7. *Delesseria sanguinea*

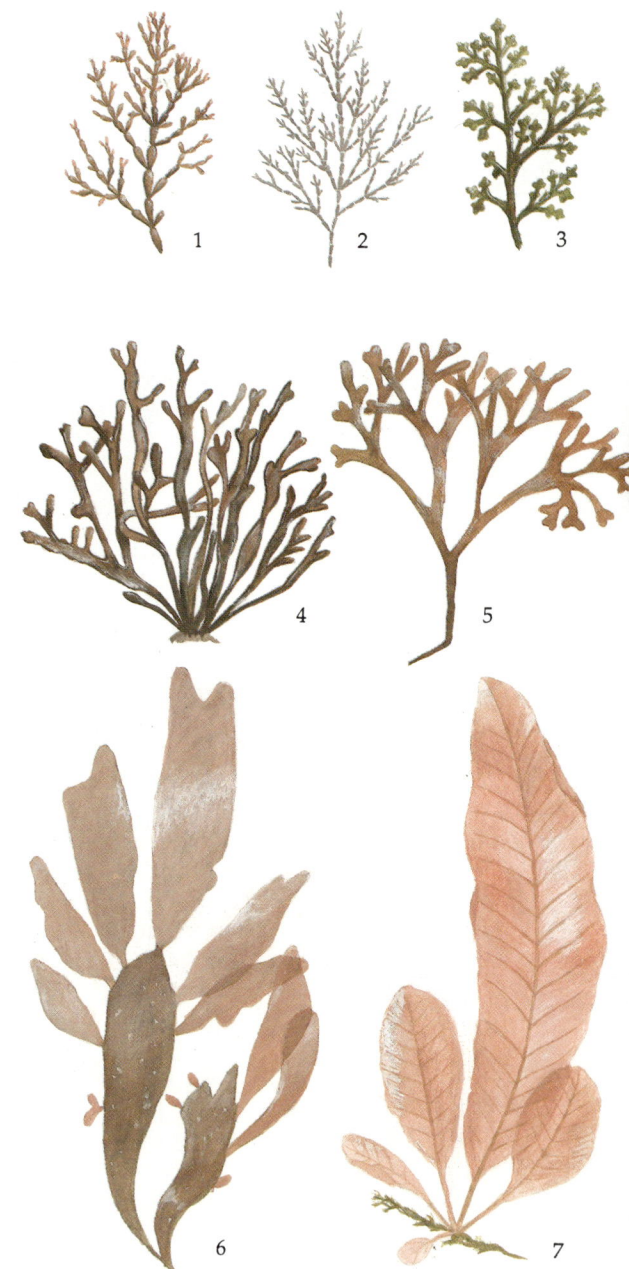

All figs. one-half natural size.

PLATE 5

Sea slugs

1. *Limapontia capitata*
2,3. *Elysia viridis*
4. *Archidoris pseudoargus*
5. *Adalaria proxima*
6. *Onchidoris muricata*

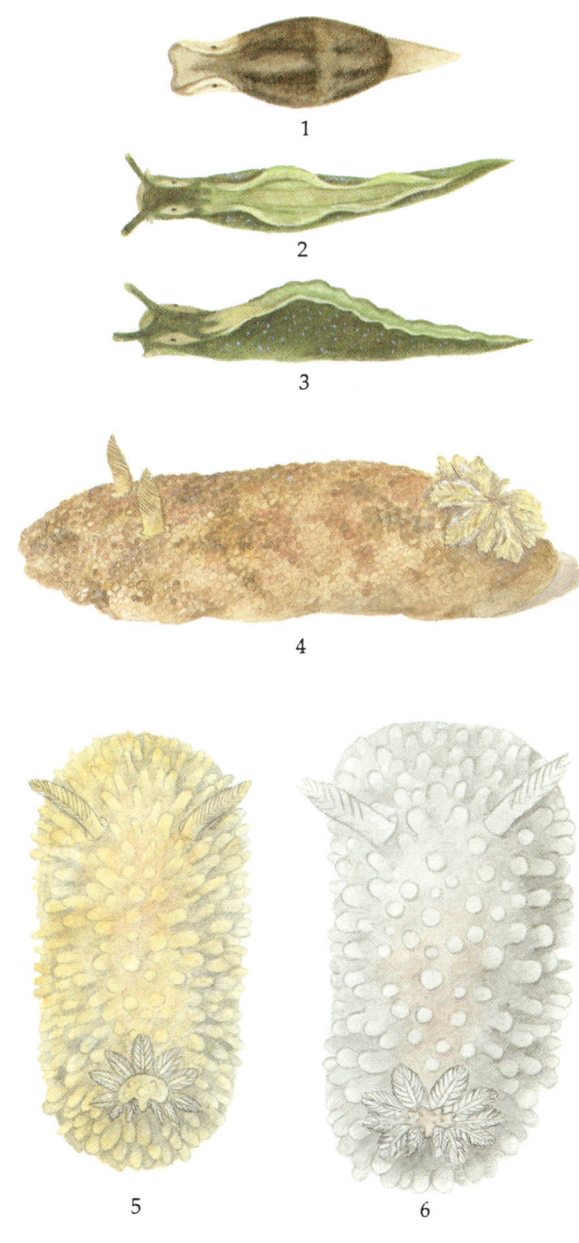

Figs. 1, 5 & 6 are one-quarter natural size.
Figs. 2-4 are natural size.

PLATE 6

Some common associations

1. The sea squirt *Botrylloides leachi* on the brown seaweed *Halidrys siliquosa*

2. The bryozoan *Flustrellidra hispida* on the red seaweed *Gigartina stellata*

3. The Star Sea Squirt *Botryllus schlosseri* on the red seaweed *Delesseria sanguinea*

4. Blue-Rayed Limpets *Patina pellucida* on the Kelp *Laminaria digitata*

Figs. 1-4 are all one-half natural size. The solitary limpet is natural size.

PLATE 7

1. *Lacuna vincta* (Montagu)
2. *Rissoa parva* (da Costa)
3. *Cerithiopsis tubercularis* (Montagu)
4. *Lacuna pallidula* (da Costa)
5. *Barleeia unifasciata* (Montagu)
6. *Tricolia pullus* (Linn.)

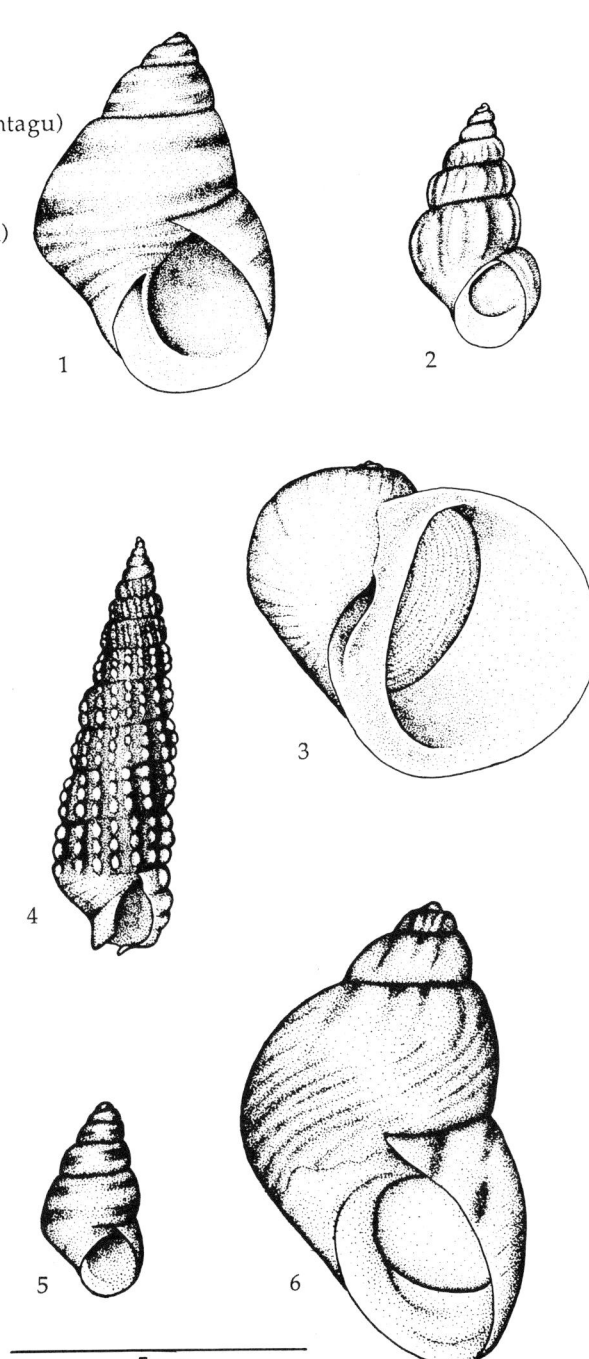

PLATE 8

1. *Cribrilina punctata* (Hassall)
2. *Bowerbankia imbricata* (Adams)
3. *Aetea anguina* (Linn.)
4. *Umbonula littoralis* Hastings
5. *Schizoporella unicornis* (Johnston in Wood)
6. *Microporella ciliata* (Pallas)

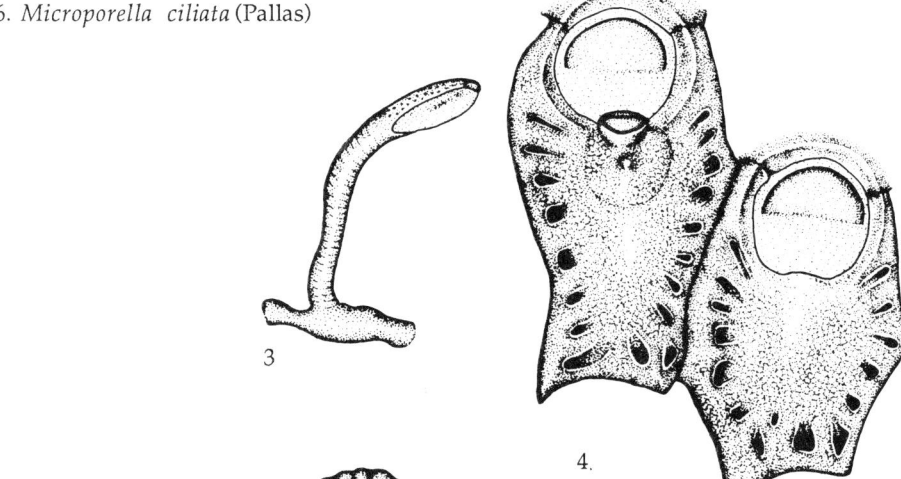

0.5 mm

II Hydroids

Hydroids, sometimes called sea-firs, may be quite common on sheltered rocky shores. Some species are particularly associated with seaweeds but many others are less selective in their choice of substrata, growing freely on large algal holdfasts and stipes, and on adjacent rock surfaces. There is no modern monograph on British hydroids, although keys, descriptions and taxonomic revisions of some families and genera have been published by Cornelius (1975, 1979, 1982).

1 Hydranths partly or wholly enclosed in a delicate chitinous cup – the hydrotheca (II.5–II.18)   4
– Hydranths not enclosed by a cup (II.1–II.3)   2

2 Colony forming a dense tuft of stalked, club-shaped hydranths, each with up to 40 thin tentacles. Pink. On brown algae, especially *Ascophyllum*   *Clava multicornis* (Forskal) (II.1)
– Colony with a slender, erect stem, branching irregularly. The hydranths spaced along the stem and branches   3

3 Outer cuticle with irregular groups of rings, separated by smooth, cylindrical portions. Up to 7 cm high; intertidal, in rock pools, frequently on algae   *Sarsia eximia* (Allman) (II.2)
– Outer cuticle regularly ringed along its whole length, typically dark brown. Intertidal, in rock pools, and often on algae   *Coryne* species (II.3)

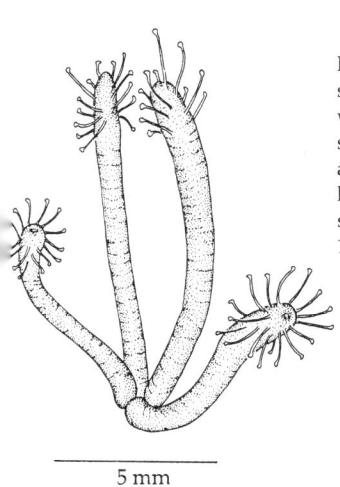

5 mm
II.1

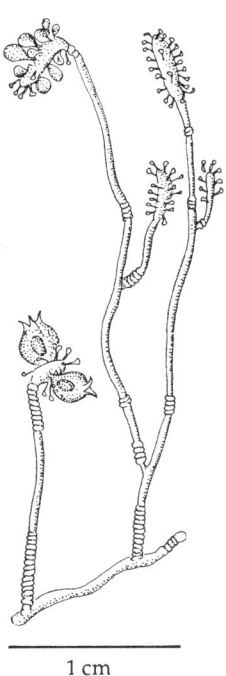

1 cm
II.2

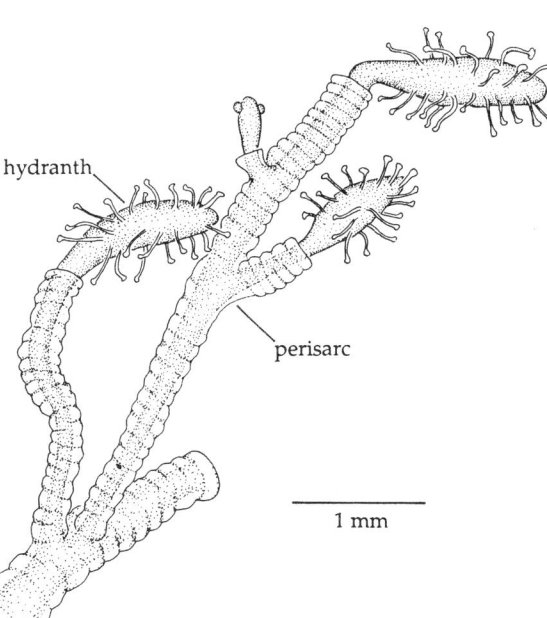

1 mm
II.3

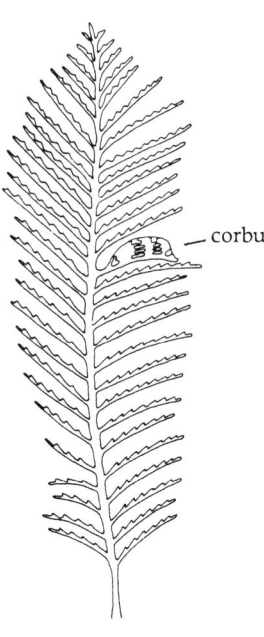

II.4

4 Colony branching regularly to form a delicate feather shape (II.4) 5
– Colony branching irregularly, not forming a feather shape 7

5 Rim of hydrotheca notched. Embryos brooded in distinctive basket-like structures (corbulae) (II.4). Up to 4 cm high. Lower shore, on various substrata, but sometimes common on *Halidrys siliquosa*, *Laminaria* holdfasts, and other large algae
    Aglaophenia pluma (Linn.) (II.4, II.5)
– Rim of hydrotheca smooth. Embryos not brooded in corbulae 6

6 Hydrotheca wide and shallow. Embryo brood chambers (gonothecae) in paired series along the main stem of colony. Large, often exceeding 10 cm long, yellowish white. Lower shore, on various substrata, including large algae
    Kirchenpaueria pinnata (Linn.) (II.6)
– Hydrotheca deeper than wide. Gonothecae situated in axils between main stem and side branches. Small, typically less than 5 cm. Lower shore, often in rock pools; on rock or large algae   *Plumularia setacea* (Ellis & Solander) (II.7)

7 Each hydrotheca attached to the stem of the colony by a short stalk (II.8–II.15) 8
– Hydrotheca not stalked, attached directly to colony stem and sometimes partially immersed in it (II.16–II.18) 14

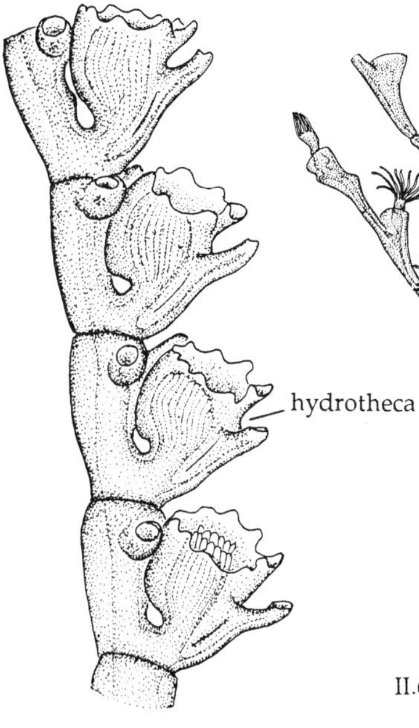

II.5   0.5 mm

II.6   1 mm

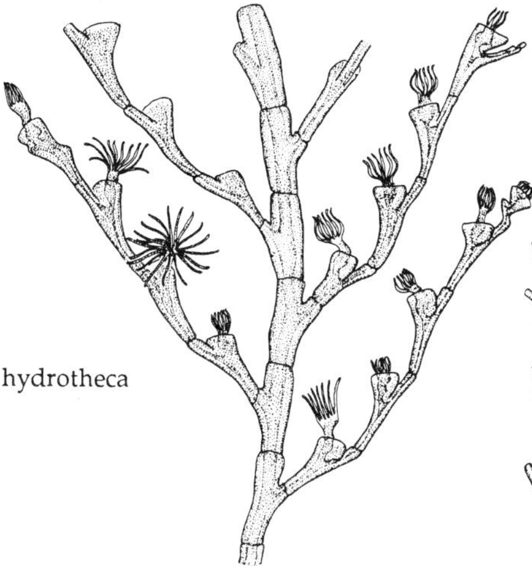

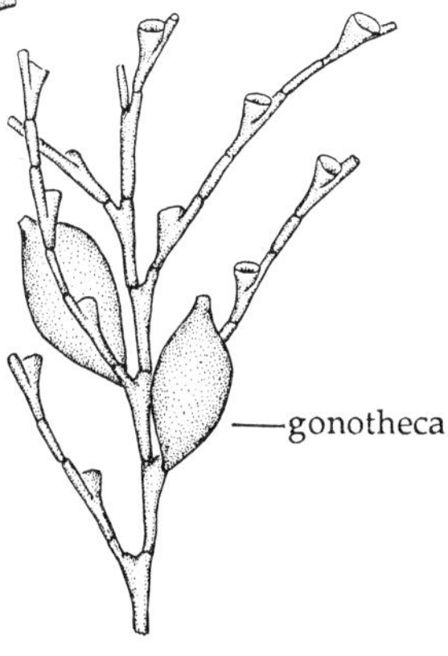

II.7   1 mm.

7 Identification

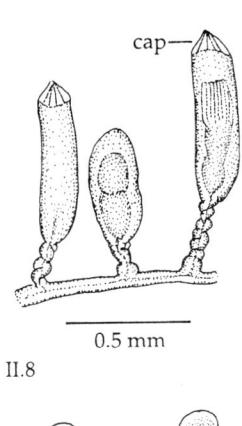

II.8

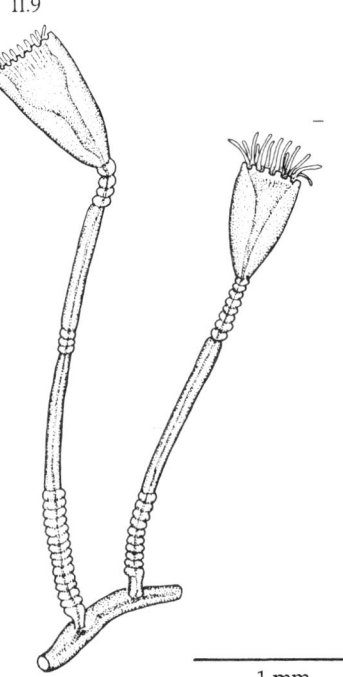

II.9

II.10

8   Colony formed of short stems, bearing very few hydrothecae, rising from an encrusting network of tubes (II.8) 9
 — Colony with larger erect stems, bearing numerous hydrothecae; often branched 11

9   Hydrotheca elongate, cylindrical; each with a conical cap at the end (lost on death of hydranth). On hydroids and algae, lower shore and subtidal   *Calycella syringa* (Linn.) (II.8)
 — Hydrotheca bell-shaped, without cap 10

10 Hydrotheca with a smooth rim, its stalk with irregular rings. On red algae, lower shore and subtidal
   Orthopyxis integra (Macgillivray) (II.9)
 — Hydrotheca with toothed rim, its stem with groups of regular rings. Common intertidally on a range of algae
   Clytia hemisphaerica (Linn.) (II.10)

11 Main stem of colony consisting of a bundle of tubes. Colony tapered, in shape rather like a small fir tree, up to 35 cm high. Hydrotheca with square-toothed rim. Lower shore and subtidal, on kelp stipes and holdfasts
   Hartlaubella gelatinosa (Pallas) (II.11)
 — Main stem of colony consisting of a single tube; colony variously shaped but not regularly tapered 12

12 Stem of colony strongly zigzagged, the outer cuticle asymmetrically thickened below each bend. Hydrotheca rather shallow, cup-shaped, with smooth rims, on short, tapered and ringed stalks. Lower shore and subtidal, often common on large brown algae, such as *Laminaria*
   Obelia geniculata (Linn). (pl. 1.5, 1.6)
 — Stem straight, curving, or sometimes slightly zigzagged, but without asymmetrical thickening of the cuticle 13

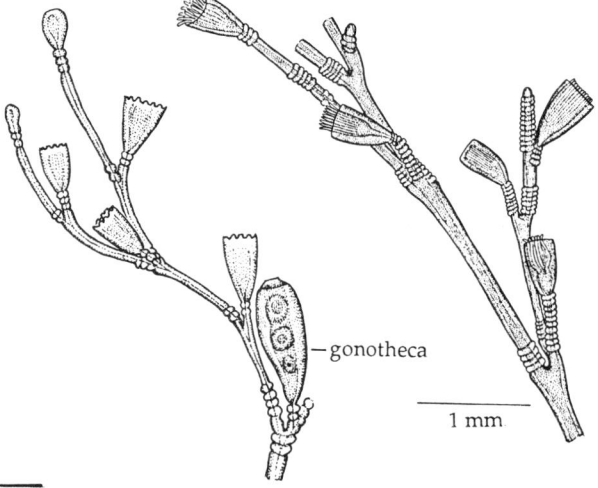

II.11 II.12

7 Identification

II.13

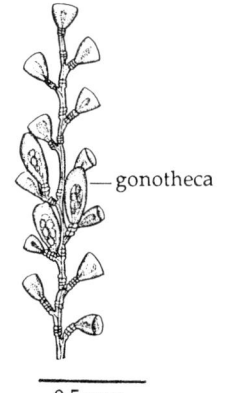

II.15

13   Stems straight, or slightly zigzagged. Hydrotheca deeply bell-shaped; rims frequently with squared teeth, but often worn smooth. Lower shore and subtidal, on various substrata, including large brown algae. Two closely similar species:
   Up to 30 cm high   *Obelia dichotoma* (Linn). (II.12)
   Up to 3 cm high   *Gonothyraea loveni* (Allman) (II.13)
   May be separated with certainty only by comparison of reproductive structures (gonothecae)
–   Stems gently curving, delicate. Hydrothecae large but rather shallow, with smooth rims. Up to 3 cm. Lower shore and subtidal, on various substrata, including algae
                                    Laomedea flexuosa Alder (II.14, II.15)

14   Hydrotheca distributed on both sides of colony branches, in more or less opposite pairs   15
–   Hydrotheca on both sides of colony branches, but not in pairs 16

15   Colony formed of short, stiff erect shoots, up to 4 cm high, sometimes branched, arising from an encrusting mass of tubes. Hydrotheca short, bent outwards from stem. Middle shore to shallow subtidal, on various substrata but often abundant on large brown algae. Particularly characteristic of *Fucus serratus*
                                    Dynamena pumila (Linn.) (pl. 1.3, 1.4)
–   Colony forming delicate, wispy tufts, up to 10 cm long, branching frequently. Hydrotheca not bent outwards from stem, its rim drawn into a fine point. Lower shore and subtidal, on brown algae, frequently on kelp stipes
                                    Amphisbetia operculata (Linn.) (II.16)

16   Colony erect, irregularly branched and tufted. Up to 3 cm. Hydrotheca smooth. Lower shore and subtidal, on a variety of substrata, including algae   *Sertularella polyzonias* (Linn.) (II.17)
–   Colony of short, straight or little-branched stems, up to 2 cm high. Hydrotheca with transverse grooves. Lower shore and subtidal, occasionally on algae   *Sertularella rugosa* (Linn.) (II.18)

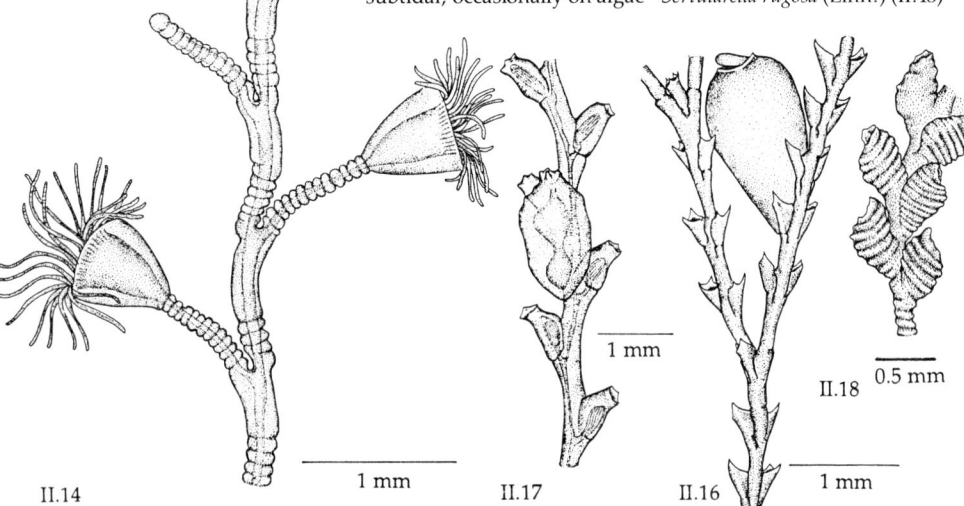

7 Identification

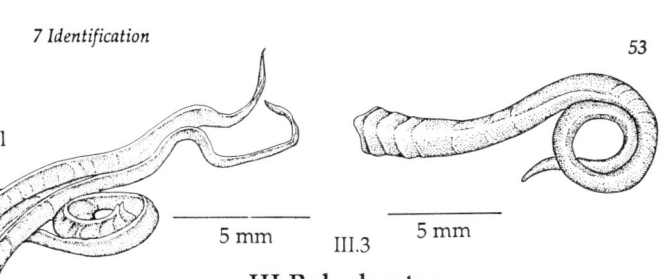

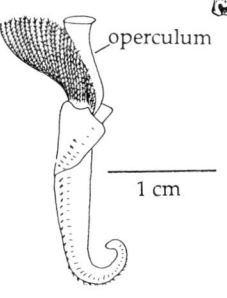

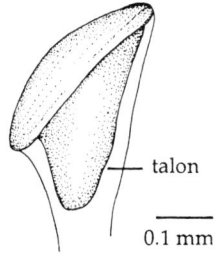

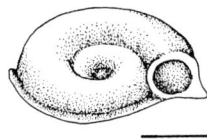

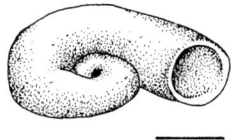

III Polychaetes

All British species are described in volumes 5 and 16 of the *Faune de France* (Fauvel, 1923, 1927). Some families have been revised more recently (Tebble & Chambers, 1982; George & Hartmann-Schroder, 1985) and all British Spirorbinae may be identified using Knight-Jones & Knight-Jones (1977). However, there is at present no complete treatise on British polychaetes.

1. Worm living in a white, calcareous tube, securely cemented to the substratum. Body with a crown of tentacles at the front end, one of which is modified as an operculum which closes the mouth of the tube (III.1)   2
– Worm free-living; or in a tube of sand or silt bonded with mucus. No operculum   10

2. Tube straight, curved or wavy, but not spiralled   3
– Tube coiled in a tight spiral   4

3. Tube with a distinct longitudinal keel, typically drawn out into a point over the mouth. Lower shore, on hard substrata, sometimes on larger kelp holdfasts. Widespread and common
                               Pomatoceros triqueter (Linn.) (III.2)
– Tube cylindrical, lacking a keel; delicate, with distinct growth rings. Usually in small, intertwined groups. Sublittoral, on hard substrata, occasionally on large holdfasts
                               Hydroides norvegica (Gunn.) (III.3)
(Stouter tubes, up to 10 cm long, with flared, upwardly directed mouth, belong to *Serpula vermicularis* Linn., a subtidal species almost entirely limited to hard substrata.)

4. Tube with a clockwise spiral   5
– Tube with an anticlockwise spiral   8

5. Operculum with a well-developed plate (talon) on one side of its stalk (III.4)   7
– Operculum without, or with only minimal development of, talon   6

6. Tube often with a flat rim where it meets substratum; body of worm pale green-brown. Characteristic on *Fucus serratus*, less frequently on *F. vesiculosus* or other large, lower shore brown algae   *Spirorbis spirorbis* (Linn.) (III.5)
– Tube without a rim; body of worm red. Lower shore, on stones, associated with the purple crustose calcareous alga *Phymatolithon*   *Spirorbis rupestris* Gee & Knight-Jones (III.6)

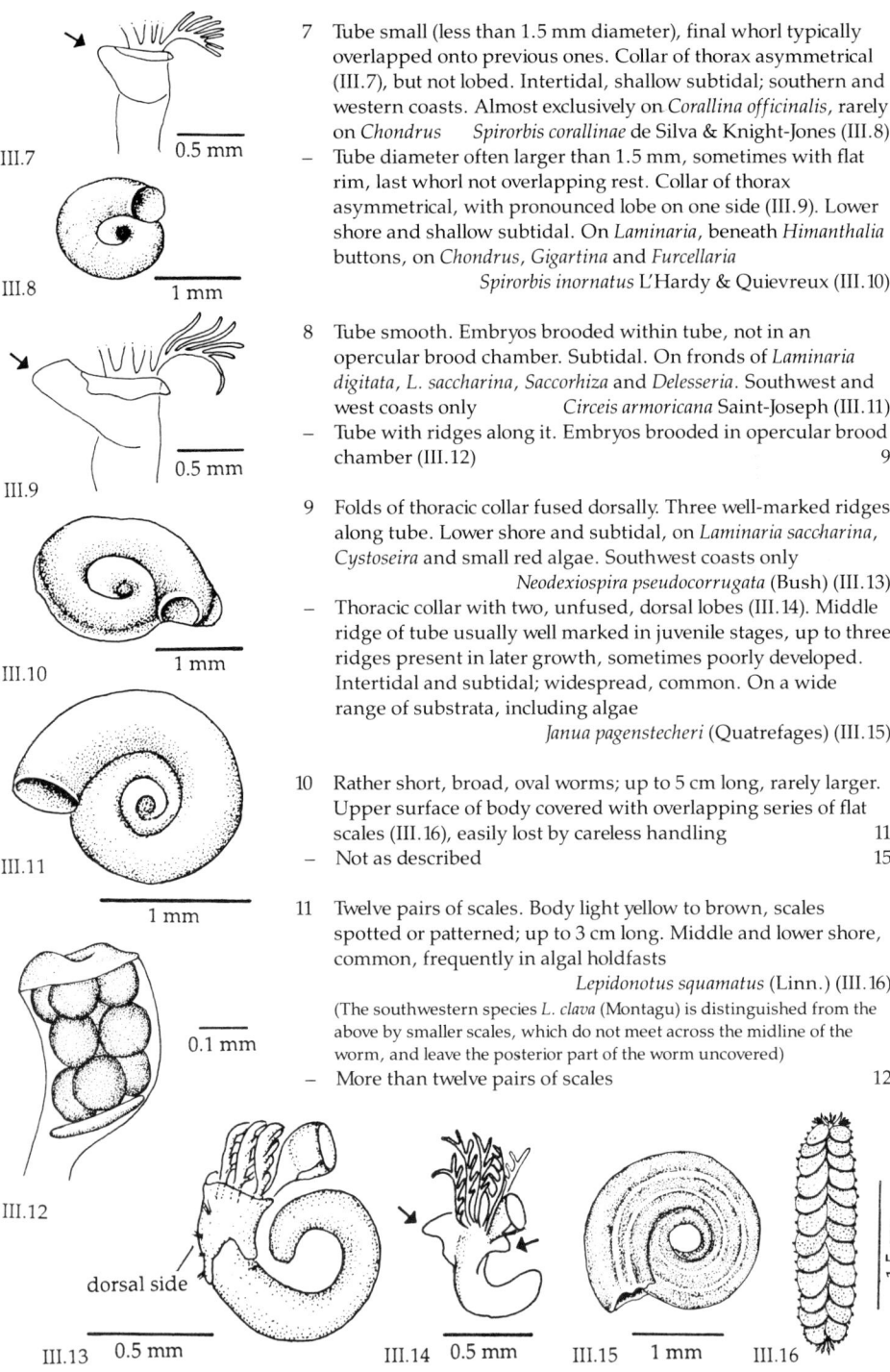

7   Tube small (less than 1.5 mm diameter), final whorl typically overlapped onto previous ones. Collar of thorax asymmetrical (III.7), but not lobed. Intertidal, shallow subtidal; southern and western coasts. Almost exclusively on *Corallina officinalis*, rarely on *Chondrus*   *Spirorbis corallinae* de Silva & Knight-Jones (III.8)
—   Tube diameter often larger than 1.5 mm, sometimes with flat rim, last whorl not overlapping rest. Collar of thorax asymmetrical, with pronounced lobe on one side (III.9). Lower shore and shallow subtidal. On *Laminaria*, beneath *Himanthalia* buttons, on *Chondrus*, *Gigartina* and *Furcellaria*
         Spirorbis inornatus L'Hardy & Quievreux (III.10)

8   Tube smooth. Embryos brooded within tube, not in an opercular brood chamber. Subtidal. On fronds of *Laminaria digitata*, *L. saccharina*, *Saccorhiza* and *Delesseria*. Southwest and west coasts only   *Circeis armoricana* Saint-Joseph (III.11)
—   Tube with ridges along it. Embryos brooded in opercular brood chamber (III.12)   9

9   Folds of thoracic collar fused dorsally. Three well-marked ridges along tube. Lower shore and subtidal, on *Laminaria saccharina*, *Cystoseira* and small red algae. Southwest coasts only
         Neodexiospira pseudocorrugata (Bush) (III.13)
—   Thoracic collar with two, unfused, dorsal lobes (III.14). Middle ridge of tube usually well marked in juvenile stages, up to three ridges present in later growth, sometimes poorly developed. Intertidal and subtidal; widespread, common. On a wide range of substrata, including algae
         Janua pagenstecheri (Quatrefages) (III.15)

10 Rather short, broad, oval worms; up to 5 cm long, rarely larger. Upper surface of body covered with overlapping series of flat scales (III.16), easily lost by careless handling   11
—   Not as described   15

11 Twelve pairs of scales. Body light yellow to brown, scales spotted or patterned; up to 3 cm long. Middle and lower shore, common, frequently in algal holdfasts
         Lepidonotus squamatus (Linn.) (III.16)
    (The southwestern species *L. clava* (Montagu) is distinguished from the above by smaller scales, which do not meet across the midline of the worm, and leave the posterior part of the worm uncovered)
—   More than twelve pairs of scales   12

7 Identification

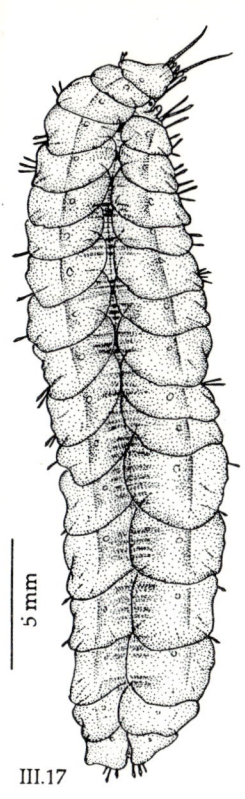

12 Eighteen pairs of transparent, gelatinous, greyish scales. Body soft, partly transparent, yellowish to brown, often larger than 5 cm. Lower shore and subtidal, occurring in *Laminaria* holdfasts   *Alentia gelatinosa* (M. Sars) (III.17)
— Fifteen pairs of scales only   13

13 Tail of worm – up to ten segments – not covered by scales, bearing two long cirri at tip (III.19, III.20)   14
— Body of worm completely covered by scales, with granular surface and fringe of fine hairs; grey, green or brownish, often spotted or flecked. Up to 3 cm long. Intertidal and shallow sublittoral, widespread and common. Often under seaweed cover, or in holdfasts   *Harmothoe impar* (Johnston) (III.18)

14 First pair of eyes close to front border of prostomium. Scales rough-surfaced, covered with small papillae, colour variable, blue, grey or brown, to red, purple or black, often patterned. Up to 7 cm long. Intertidal and sublittoral, frequently in holdfasts   *Harmothoe imbricata* (Linn.) (III.19)
— First pair of eyes distant from front border of prostomium. Scales grey, brown or reddish, with a clear central area. Up to 7 cm long. Lower shore, widespread and common, occurring in *Laminaria* holdfasts   *Harmothoe extenuata* (Grube) (III.20)

15 Front end of worm with a tangled mass of threadlike tentacles, or a more regular fan of stiff, feathery ones. Living in tube: leathery, with sand, gravel and shell attached, or formed from mucus-agglutinated sand and silt   16
— Front end of worm with various paired appendages but lacking a tentacle crown, head usually distinct   19

III.17

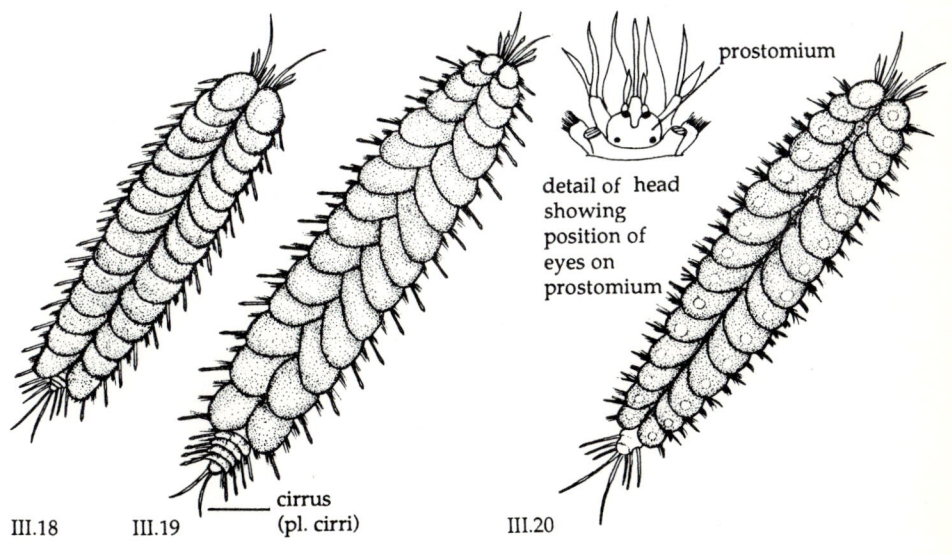

detail of head showing position of eyes on prostomium

prostomium

cirrus (pl. cirri)

III.18   III.19   III.20

16 Front end of worm with mass of threadlike tentacles (III.21). Tube delicate, of sand and silt bound with mucus 17
 – Front end of worm with stiff, regular tentacle fan. Tube tough, leathery 18

17 Three pairs of bright red branching gills behind tentacles. Worms rather fat, up to 30 cm long, but usually much shorter. Building insubstantial tubes; often in large kelp holdfasts, on sheltered shores with adjacent sand   *Amphitrite* species (III.21)
 – Two pairs of red gills behind tentacle crown. Up to 6 cm long. Tube of sand and detritus, attached to algae, hydroids, sponges and bryozoan turfs. Lower shore   *Nicolea* species (III.22)

18 Tentacles rather short and stout. Each tentacle with paired appendages. Up to 5 cm long. Lower shore and shallow subtidal, often in kelp holdfasts
    Branchiomma bombyx (Dalyell) (III.23)
 – Tentacles long and slender, without appendages, but with distinct eyespots. Up to 15 cm long. Lower shore and shallow subtidal, south and west coasts. Often in large kelp holdfasts
    Megalomma vesiculosum (Montagu) (III.24)

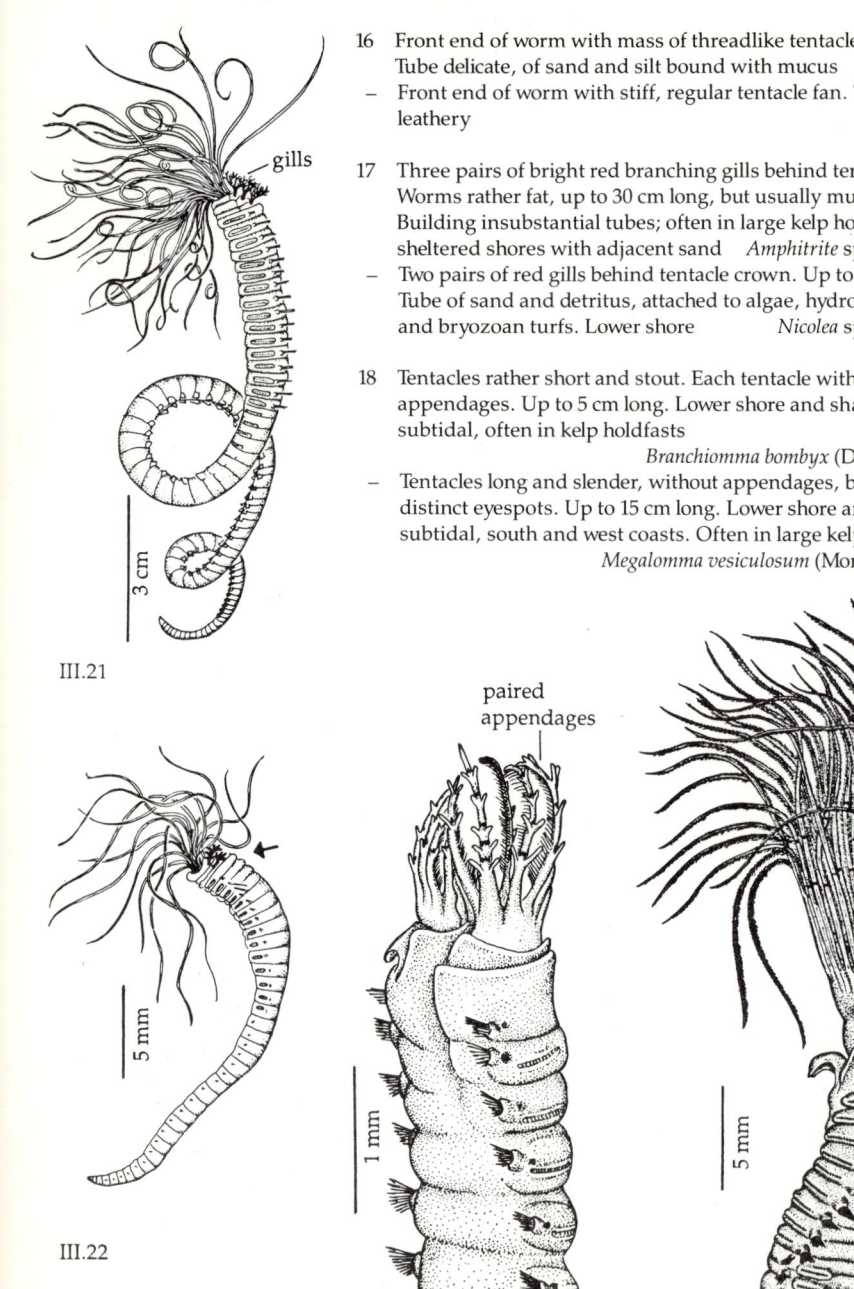

III.21 III.22 III.23 III.24

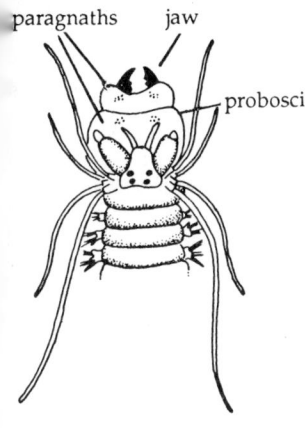

paragnaths   jaw
proboscis

III.25

19 Worms with well-developed heads bearing paired antennae, paired palps and several pairs of tentacle-like cirri; eversible proboscis† with large, powerful jaws (III.25) 20

†Gentle pressure with finger, applied just behind the head of the worm, will cause the proboscis to evert.

– Not as described 22

20 Proboscis, when everted, with distinct groups or lines of small black teeth (paragnaths) (III.26); cirri rather short 21

– Proboscis with very small, poorly developed paragnaths. Cirri very long, reaching back to at least the tenth body segment. Up to 6 cm long, slender. Lower shore and shallow subtidal, in *Laminaria* holdfasts, where it may secrete mucus tubes
Platynereis dumerilii (Aud. & M. Edwards) (III.25)

21 Paragnaths in two small groups. Up to 12 cm long; colour variable, often greenish-bronze. Lower shore, among algae and in holdfasts *Nereis pelagica* (Linn.) (III.26)

– Paragnaths in distinct transverse series (III.27). Up to 25 cm long; greenish-bronze, with red tints. Lower shore, among algae and in holdfasts *Perinereis cultrifera* (Grube) (III.28)

22 Head of worm bearing three antennae, two short palps and one or two pairs of tentacle-like cirri (III.29) 23

– Head of worm with two, four or five antennae 25

23 Head with palps clearly separated. Each body segment with a pair of elongate, jointed cirri. Delicate slender worms, up to 3 cm long, often common in algal holdfasts *Syllis* species (III.29)

– Head with palps fused at base, or for whole length; cirri on body segments smooth or only indistinctly jointed 24

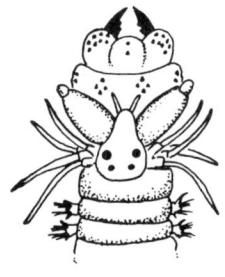

III.26

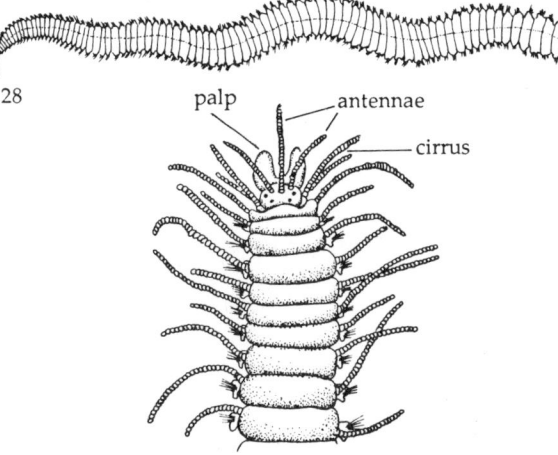

III.28

palp   antennae
cirrus

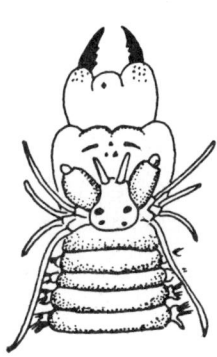

III.27

III.29

III.30

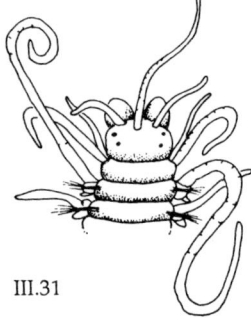

III.31

24 Head with palps fused at base. Small, fragile worms, often common in algal turfs and among holdfasts. Numerous species and genera, including: *Odontosyllis* (III.30)
Eusyllis (III.31)
– Head with palps fused for whole of length. Numerous species of small (less than 2 cm) or very small (2–4 mm) worms, often common in algal turfs and among holdfasts. Genera include:
Grubea (III.32)
Sphaerosyllis (III.33)

25 Head of worm with one or two pairs of antennae and two or three pairs of long tentacle-like cirri (III.34). Body appendages (parapodia) each bearing a long cirrus (III.35) 26
– Head of worm with four or five antennae and four pairs of rather short tentacle-like cirri. Rest of body with paddle-like cirri (III.36) 27

26 Head with one pair of antennae and three pairs of tentacle-like cirri. Up to 3 cm long, yellowish with dark banding. Lower shore and subtidal, often in *Laminaria* holdfasts
Castalia punctata (O. F. Muller) (III.35)
– Head with two pairs of antennae and two pairs of tentacle-like cirri. Up to 8 cm long; yellow, brown or reddish. Lower shore, often in *Laminaria* holdfasts
Kefersteinia cirrata (Keferstein) (III.34)

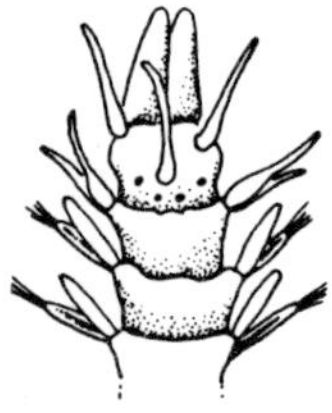

III.32

III.34

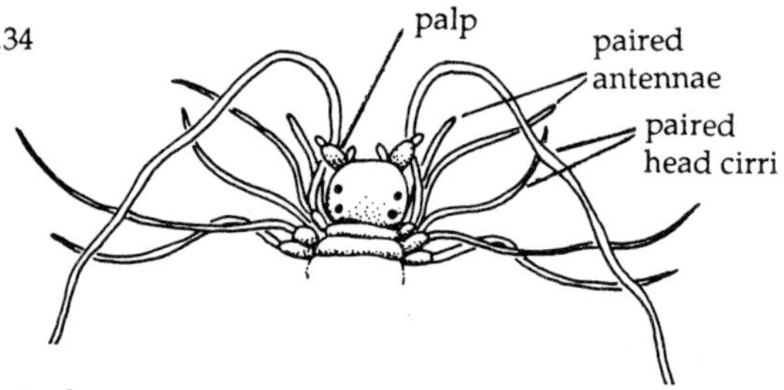

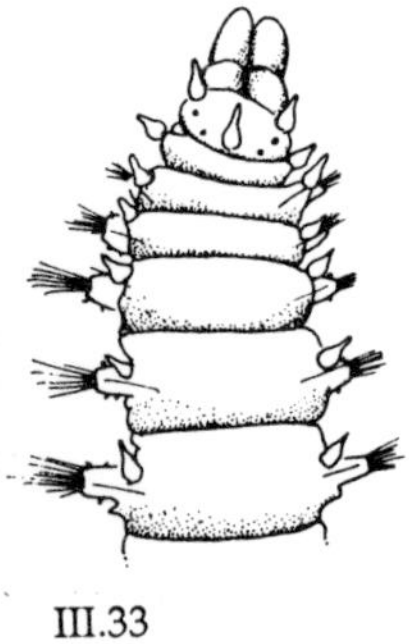

III.33

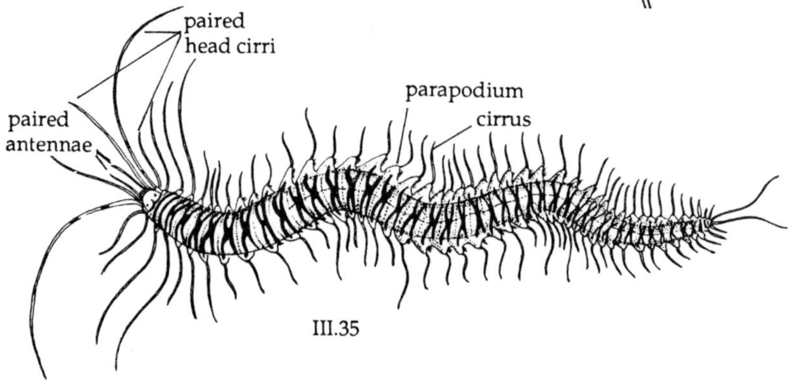

III.35

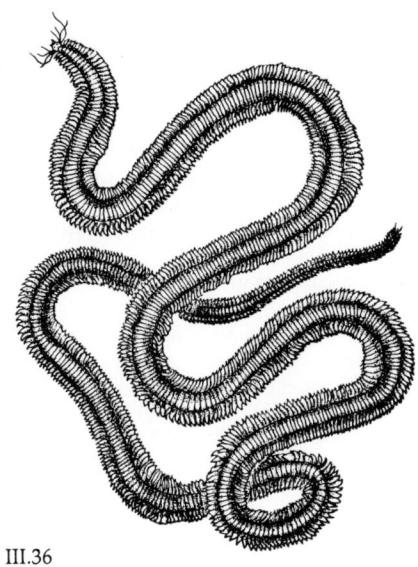

III.36

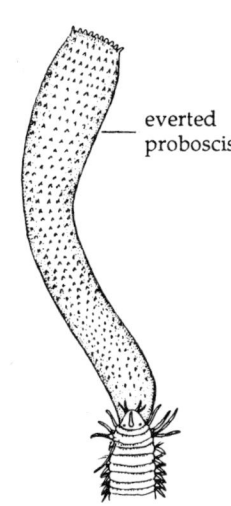

everted proboscis

III.37

27 Head elongate, with four short antennae and four pairs of cirri. Body with conspicuous paddle-like cirri; iridescent bluish-green, often very large (up to 60 cm long). A lower shore species sometimes occurring in larger holdfasts
Phyllodoce lamelligera (Gmelin) (III.36)

– Head short (III.37) with four antennae at front and an extra one on top of head 28

28 Head rounded. Up to 15 cm long, slender, brilliant emerald green. Middle and lower shore, often under algae and in holdfasts. Widespread and common
Eulalia viridis (O. F. Muller) (III.37, III.38)

– Head heart-shaped. Up to 6 cm long, slender, brown or red. Lower shore, often in *Laminaria* holdfasts
Eulalia sanguinea Oersted (III.39)

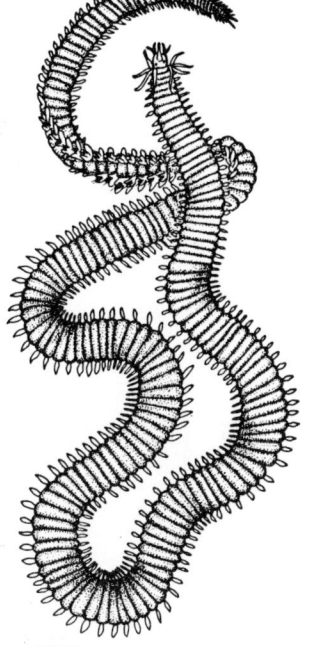

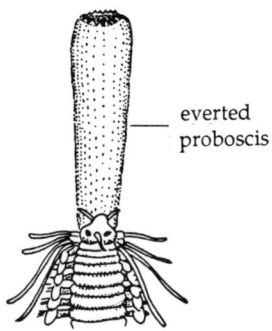

everted proboscis

III.38

III.39

IV Isopods

All British species of Isopoda, including the species keyed here, are described and illustrated by Naylor (1972).

1 Adults with five pairs of pereopods. Pleon forming a narrow segmented 'tail'. Males, females and juveniles dissimilar. A common crevice-dwelling species, often occurring in *Laminaria* holdfasts   *Gnathia maxillaris* (Montagu) (IV.1)
-   Adults with seven pairs of pereopods. Body not as described   2

2 Uropods attached to the sides of, or beneath, the end portion of the body (IV.2)   3
-   Uropods attached at the very end of the body (IV.11)   12

3 Uropods at the sides, flattened and forming a tail fan with the telson (IV.2), or elongate and projecting   4
-   Uropods beneath the body, not visible from above; hinged to form flaps covering the pleopods (IV.6)   6

4 Elongate, almost cylindrical animals with prominent eyes. First pair of pereopods modified to form gnathopods, often held below body. Uropods two-branched, broad, forming a marked tail fan. Females up to 11 mm long, males 4 mm. A crevice-dwelling species occurring occasionally in kelp holdfasts   *Anthura gracilis* (Montagu) (IV.2)
-   Short, broad animals, strongly convex above, males with paired or single dorsal processes. Uropods two-branched or one-branched, narrow, projecting beyond end of body   5

5 Uropods one-branched, elongate, projecting conspicuously. Male with a single dorsal process on pereon segment 6. Rolls into a ball when disturbed. Up to 4 mm long. An upper shore species often common in *Lichina*
        Campecopea hirsuta (Montagu) (IV.3)
-   Uropods two-branched. Male with paired processes on pereon 6 and a blunt tubercle behind. Up to 7 mm long. A crevice-dwelling species often common in kelp holdfasts and among lower shore algae   *Dynamene bidentata* (Adams) (IV.4)

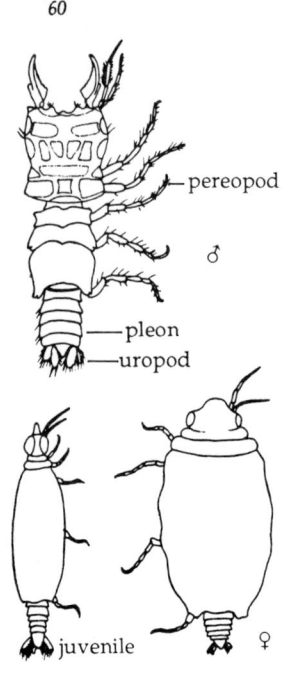

IV.1

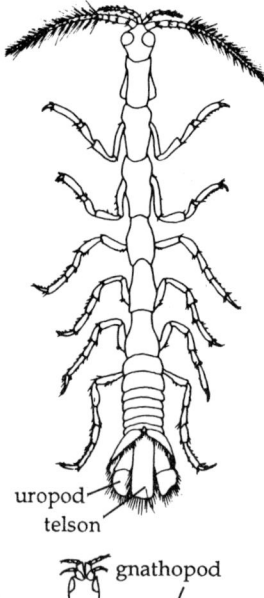

IV.2

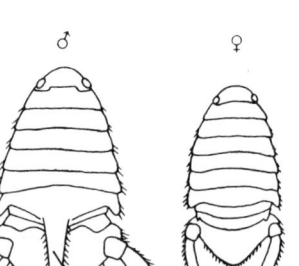

IV.3

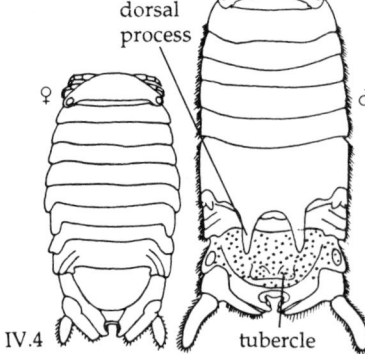

IV.4

7 Identification

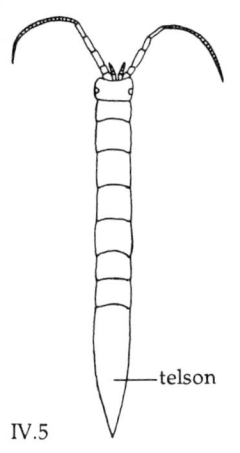

IV.5

pereon

pleon

telson

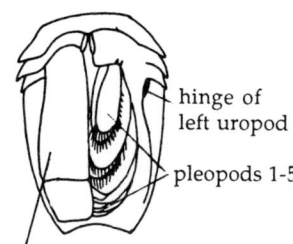

hinge of left uropod

pleopods 1–5

uropod   2 mm

pleotelson in ventral view; left uropod removed.

IV.6

6 Body with seven distinct segments behind the head. Remaining segments fused with the elongate, pointed telson. On *Halidrys siliquosa*; up to 25 mm long, resembling the bladders of the alga. South and west coasts
  Synisoma acuminatum Leach (IV.5)
– Body with seven pereon segments and two pleon segments (IV.6) 7

7 Hind margin of telson straight or concave. Males up to 30 mm long, females up to 18 mm; light or dark brown with white markings. On detached algae and drift weed, or sometimes among bushy fucoids *Idotea emarginata* (Fabricius) (IV.6)
– Hind margin of telson pointed or tapered 8

8 Hind margin of telson more or less three-toothed. Males up to 30 mm long, females up to 18 mm; green or brown with white spots or lines. On detached algae and drift weed, often common among bushy fucoids *Idotea baltica* (Pallas) (pl. 2.6)
– Hind margin of telson pointed or rounded, but not three-toothed 9

9 Hind margin of telson with a conspicuous point; slightly convex at the sides. Males up to 20 mm long, females up to 13 mm; uniformly red, brown or green, sometimes with longitudinal white marks. Common in *Ascophyllum* and *Fucus* on sheltered rocky shores, juveniles in *Cladophora*, *Polysiphonia* and other small algae *Idotea granulosa* Rathke (IV.7)
– Margin of telson rounded or with only a blunt point; straight or slightly convex at the sides 10

10 Flagellum of antenna much shorter than peduncle, with dense, fine bristles in males. Up to 11 mm long; dark purplish brown with white markings. Among fucoid algae on exposed rocky shores *Idotea pelagica* Leach (IV.8)
– Flagellum of antenna longer than peduncle, without dense bristles 11

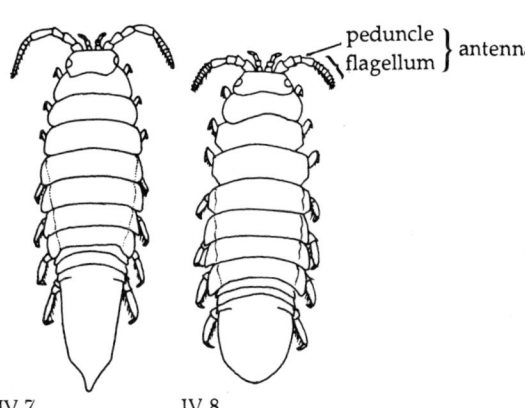

peduncle
flagellum } antenna

IV.7   IV.8

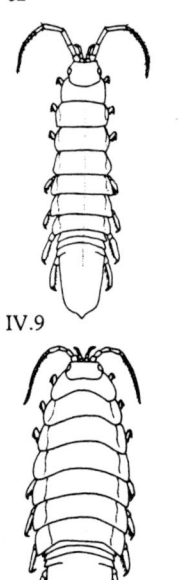

IV.9

IV.10

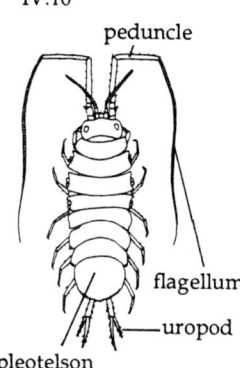

peduncle

flagellum

uropod

pleotelson

peduncle } antenna
flagellum

IV.11

7 Identification

11 Body slender, four to five times as long as wide. Males up to 15 mm long, females up to 10 mm; green or brown. In estuaries among lower shore algae, or in brackish high shore pools
Idotea chelipes (Pallas) (IV.9)

– Body broad, little more than three times as long as wide. Males up to 30 mm long, females up to 16 mm; brown, with white streaks or blotches. On lower shore among drift algae, often with *I. emarginata* and *I. baltica*   *Idotea neglecta* Sars (IV.10)

12 Body oval, head large with eyes borne on lobes at the side. Antennae as long as body, flagellum shorter than peduncle (labelled on IV.11). Pereopods 2 – 7 mm long and slender, exceeding body length. A small, delicate animal, up to 3 mm long, appearing rather spider-like. On kelp holdfasts and among erect bryozoans and hydroids on the lower shore
Munna kroyeri Goodsir (pl. 2.5)

– Body oval or elongate, eyes on upper surface of head. Antennal flagellum longer than peduncle. Pereopods shorter than body   13

13 Antennae longer than body. Uropods elongate, longer than pleotelson, projecting conspicuously. Up to 10 mm long. On kelp holdfasts, and on encrusting bryozoans, hydroids and sponges   *Janira maculosa* Leach (IV.11)

– Antennae shorter than body. Uropods tiny, accommodated within a notch in the hind margin of the pleotelson. Intertidal among fucoid algae   *Jaera* species (pl. 2.7, 2.8)

V Gammaridean amphipods

There are more than 250 British species of Gammaridea. Identification need not be difficult; the initially intimidating terminology refers precisely to discrete morphological features which are easily seen with a minimum of preparation. The main features of the amphipod body are shown in Fig.V. 0, and characters used in this key are shown and labelled in the accompanying figures. The male (♂) usually differs from the female (♀), which is most easily recognised by the presence of eggs beneath the body, or by the plates of the brood chamber between the front series of legs. All British species are admirably described and illustrated in Lincoln (1979). The following key, based largely on Lincoln's more extensive keys, will allow identification *only* of those species most commonly associated with seaweeds. Species not accurately characterised by this key should be identified using Lincoln (1979).

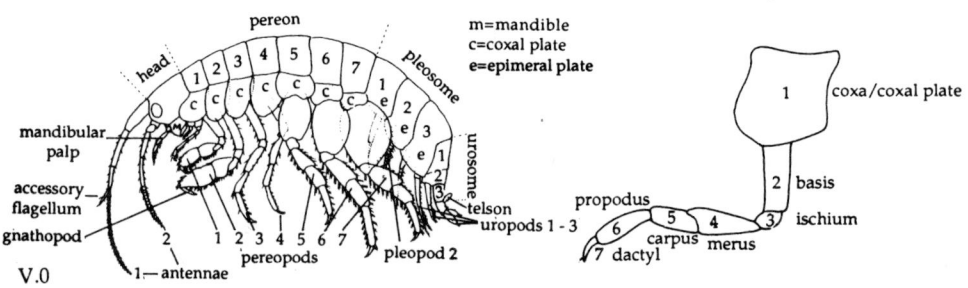

V.0

7 Identification

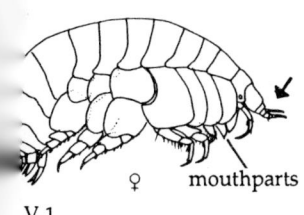

V.1

1. Gnathopod 2 long and slender, with minute dactyl. Antenna 1 with distinct, broad basal segment and short second and third segments; accessory flagellum conspicuous (V.1) 2
  – Not with the above association of characters 4

2. Head with piercing mouthparts, formed into a conical bundle; eyes small and round. Up to 9 mm long; pinkish red. In kelp holdfasts .... *Acidostoma sarsi* Lincoln (V.1)
  – Mouthparts not as described 3

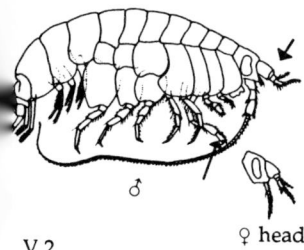

V.2

3. Telson with a small split extending about one-third of its length. Gnathopod 2 with long, slender propodus. Head with large oval eyes. Up to 8 mm long; pale yellow or cream, sometimes with red flecks. In kelp holdfasts
  .... *Orchomene humilis* (Costa) (V.2)
  – Telson with elongate split extending more than half its length. Gnathopod 2 with short, fat propodus. Head with large oval eyes. Up to 5 mm long; greyish or white. Common on all coasts; often in kelp holdfasts *Orchomene nana* (Kroyer) (V.3)

4. First coxal plate small, partly or completely hidden by second; plates 2–4 very large (V.4). No accessory flagellum. Telson undivided 5
  – Not with the above association of characters 8

5. Uropod 3 two-branched, each branch with one segment only (V.4) 6
  – Uropod 3 unbranched, with two segments (V.6) 7

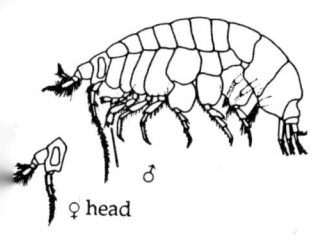

V.3

6. Head with large, curved rostrum; antennae short, of equal length. Mandible with small molar. Up to 5 mm long; pale brown or red. Lower shore, in kelp holdfasts
  .... *Amphilochus manudens* Bate (V.4)
  – Head with short, curved rostrum; antenna 2 longer than 1. Mandible with prominent ridged molar. Up to 3 mm long; broadly banded with dark brown or black/violet. All coasts, common. In kelp holdfasts and among fucoids
  .... *Gitana sarsi* Boeck (V.5)

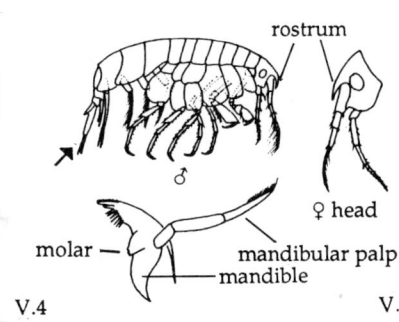

V.4

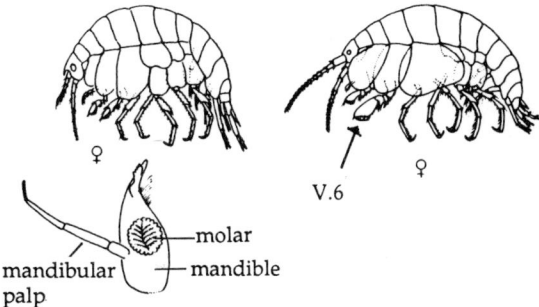

V.5   V.6

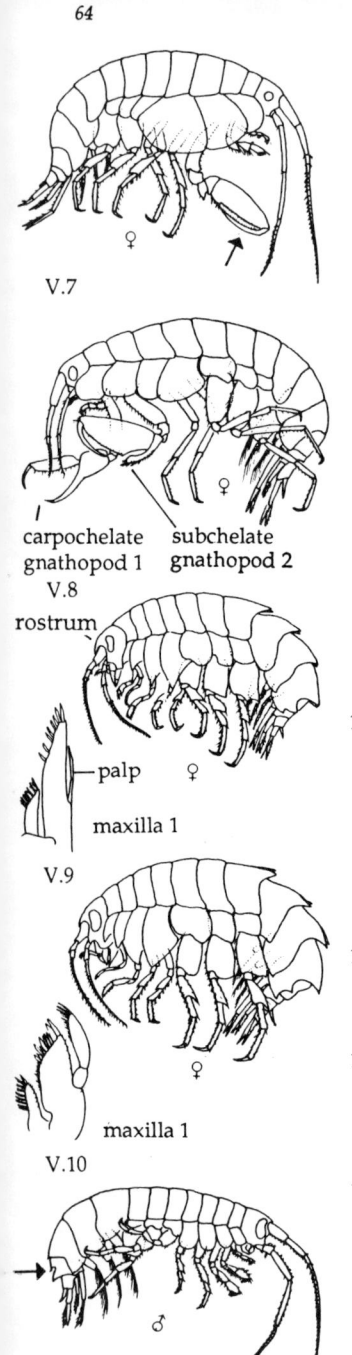

7   Gnathopod 2 with rectangular propodus, palm (surface facing dactyl) almost perpendicular to long axis. Eyes small, round, red. Up to 3 mm long; white with red flecks. All coasts, common. In kelp holdfasts and among lower shore fucoids
Stenothoe monoculoides (Montagu) (V.6)
–   Gnathopod 2 with elongate oval propodus, palm almost parallel to long axis. Eyes large, round, dark red. Up to 6 mm long; white with yellow and pink markings. All coasts, common. In kelp holdfasts   *Stenothoe marina* (Bate) V.7

8   Gnathopod 1 carpochelate; gnathopod 2 larger than 1, massive, subchelate. Eyes large, round, red. Up to 18 mm long; pinkish with dark bands, sometimes pale green above. All coasts, often common. Lower shore among fucoid algae, in kelp holdfasts, often associated with sponges and ascidians
Leucothoe spinicarpa (Abildgaard) (V.8)
(*L. incisa* Robertson is smaller than the above species – up to 7 mm – and differs also in having a much smaller dactyl to gnathopod 1, less than one-quarter length of propodus.)
–   Not with the above association of characters   9

9   Head with large curved rostrum and prominent eyes. Piercing mouthparts forming a conical bundle; mandible slender (V.9) 10
–   Not with the above association of characters   11

10 Maxilla 1 with very small palp. (Use very fine forceps to pull out the mouthparts.) Up to 6 mm long; yellowish with dark red or brown bands. Widespread and common. Lower shore, among algae, often in kelp holdfasts
Iphimedia minuta (Sars) (V.9)
–   Maxilla 1 with well-developed palp. Up to 12 mm long; white or yellowish, with red or pink bands. All coasts. Lower shore, often in kelp holdfasts   *Iphimedia obesa* Rathke (V.10)

11 Mandible with a palp, visible between or below bases of antennae (V.12)   12
–   Mandible without a palp   36

12 Telson elongate, flat, undivided or with a cleft in the middle. Fourth coxal plate concave behind   13
–   Telson short and thick, undivided or with a small notch at tip. Fourth coxal plate not concave behind   21

13 Antenna 1 without, or with a very small and indistinct, accessory flagellum   14
–   Antenna 1 with a conspicuous accessory flagellum   16

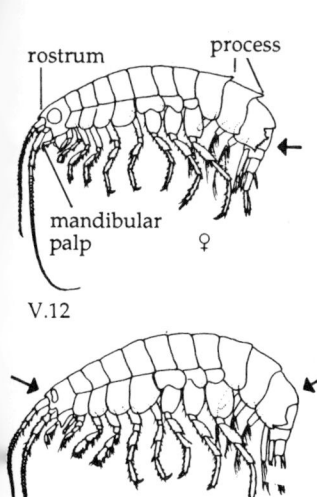

V.12

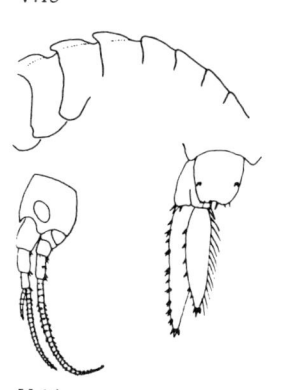

V.13

V.14

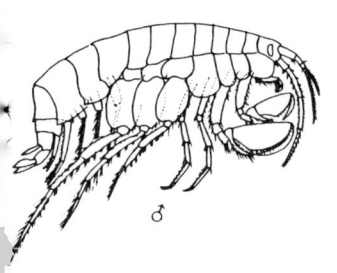

V.15

14 Telson cleft. Segments 2 and 3 of urosome fused, segment 1 with small bump and prominent pointed tooth. Eyes large. Up to 10 mm; white, with brown patches. Widespread and common. Lower shore, among kelp holdfasts
Atylus swammerdami (Milne Edwards) (V.11)

– Telson undivided, or with only small notch. Urosome segments not fused 15

15 Processes on pleon segments 1 and 2. Head with prominent rostrum and large round eyes. Epimeral plate 3 strongly toothed on hind margin. Up to 7 mm long; white to purplish, often with dark blotches. Widespread and common. Lower shore, in kelp holdfasts and *Corallina* turf
Apherusa bispinosa (Bate) (V.12)

– No processes. Head with large, kidney-shaped eyes. Up to 8 mm long; yellowish white, pink, or pinkish purple with white dorsal patch, with spots or blotches of bright red-brown. All coasts, common. Lower shore, among algae and in kelp holdfasts *Apherusa jurinei* (Milne Edwards) (V.13)

16 Body with well-marked sharp ridge along the back. Antennae 1 and 2 of equal length. Eyes large. Gnathopods 1 and 2 about equal size. Up to 15 mm long; yellowish with red-brown blotches. Widespread and common. Intertidal and sublittoral, in kelp holdfasts and red algal turfs
Gammarellus angulosus (Rathke) (V.14)

(*G. homari* (Fabricius) is a larger species – up to 35 mm long – with small eyes, and the ridge strongly extended behind. It has been found in *Corallina* turf from eastern coasts only.)

– Not with the above association of characters 17

17 Coxal plates large, overlapping. Antenna 1 shorter than 2; accessory flagellum about half length of antenna 1 flagellum. Gnathopods 1 and 2 large, subchelate. Pereopods 3 and 4 very slender. Eyes large, oval. Up to 10 mm long; pale orange, or white with red or orange patch. West coasts only. Lower shore, in kelp holdfasts *Liljeborgia pallida* (Bate) (V.15)

– Not with the above association of characters 18

18 Gnathopods 1 and 2 subchelate, about equal size. Upper surface of urosome with groups of short spines. Inner branch of uropod 3 very small. Eyes very large, kidney-shaped. Up to 25 mm long; dark green, sometimes with red or yellow patches. Widespread and common. Intertidal, associated with a wide range of algae *Chaetogammarus marinus* (Leach) (pl. 2.4)

– Gnathopod 2 much larger than 1. Urosome spines sparse or absent 19

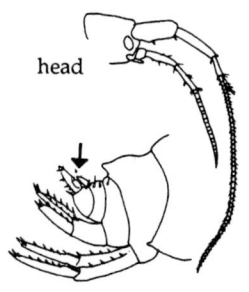
head

urosome

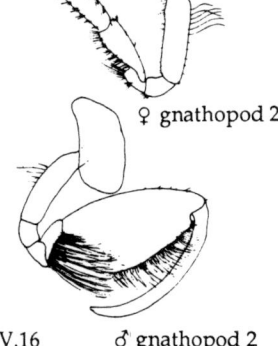
♀ gnathopod 2

V.16   ♂ gnathopod 2

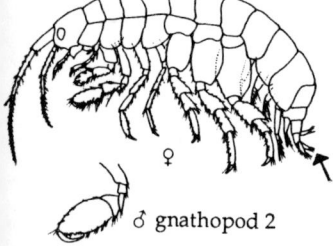
♀

♂ gnathopod 2

V.17

19   Outer branch of uropod 3 very short. Urosome with ridge along the top. Eyes small and round. Up to 10 mm long; yellow-brown. Southwest and west coasts. Intertidal and sublittoral, among fucoids and in kelp holdfasts
Gammarella fucicola (Leach) (V.16)
–   Uropod 3 with both branches of about equal length   20

20   Branches of uropod 3 short and fat, about as long as peduncle (stalk). Telson deeply divided; each lobe rounded at the tip, with small groups of spines just in front of the tip. Up to 10 mm long; yellow-white flecked with violet, or violet with white patches. Southwest and west coasts. Lower shore, among fucoids and in kelp holdfasts   *Elasmopus rapax* Costa (V.17)
–   Branches of uropod 3 very elongate. Telson widely cleft, with spines at the tip of each lobe. Antenna 1 shorter than 2. Gnathopods 1 and 2 simple in female; gnathopod 1 simple, gnathopod 2 subchelate in male. Up to 10 mm long; yellow or orange, with red blotches. All coasts. Lower shore, in kelp holdfasts   *Cheirocratus sundevallii* (Rathke) (V.18)

21   Uropods 1 and 2 distinct, spiny; uropod 3 reduced to inconspicuous plate beside telson. Segment 1 of urosome more than twice length of segment 2. Up to 4 mm long; dark red or brown-red, sometimes with purple patch. Southwest and west coasts. Intertidal and shallow subtidal, in kelp holdfasts and *Corallina* turf   *Podocerus variegatus* Leach (V.19)
–   Urosome with 3 pairs of uropods   22

22   Antenna 2 longer than 1, with massively developed peduncle (V.20). Urosome flattened. Uropods 1 and 2 two-branched; uropod 3 short, unbranched   23
–   Not with the above association of characters   25

23   Urosome segments fused; distinct ridges at the sides hiding insertion of uropods when seen from above. Antenna 1 of female with four large teeth on lower margin of first segment; antenna 2 of male with one large and one small tooth on tip of lower edge of fourth segment and a single tooth near base of fifth segment. Up to 4 mm long. Southwest only, subtidal. Builds tube among algae, hydroids and sponges. Locally abundant   *Corophium acutum* Chevreux (V.20)
–   Urosome segments fused, but no ridges at the sides; uropods inserted in lateral notches, visible from above   24

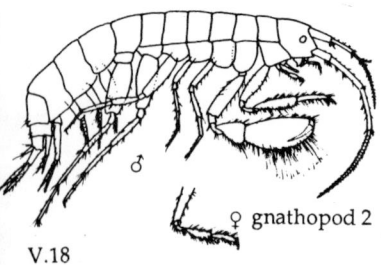
♂

♀ gnathopod 2

V.18

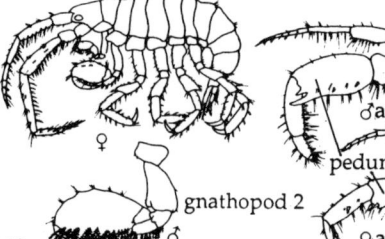
♀

gnathopod 2

♂

V.19

♂ antennae
peduncle
urosome
♀ antennae
V.20

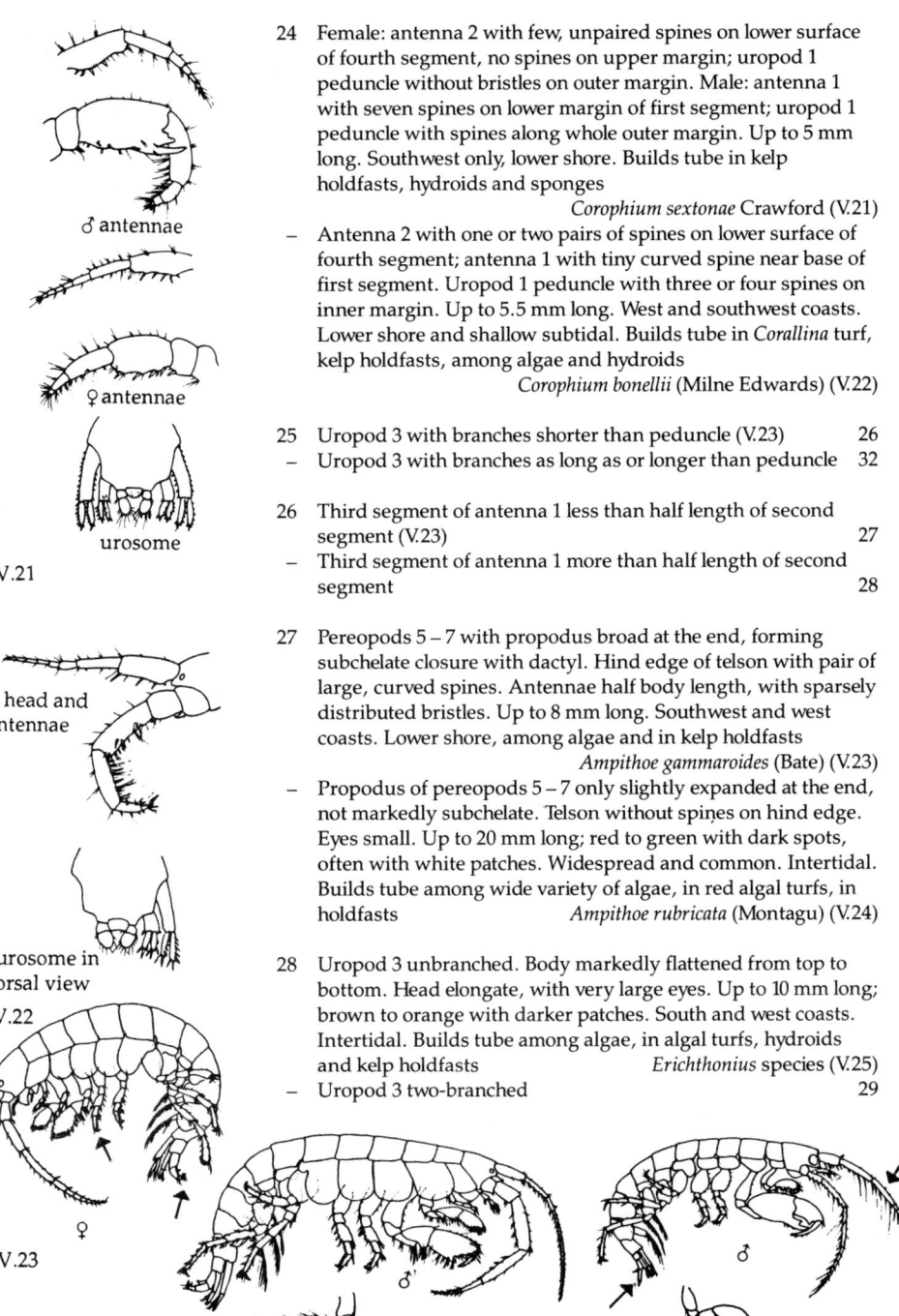

24 Female: antenna 2 with few, unpaired spines on lower surface of fourth segment, no spines on upper margin; uropod 1 peduncle without bristles on outer margin. Male: antenna 1 with seven spines on lower margin of first segment; uropod 1 peduncle with spines along whole outer margin. Up to 5 mm long. Southwest only, lower shore. Builds tube in kelp holdfasts, hydroids and sponges
Corophium sextonae Crawford (V.21)

– Antenna 2 with one or two pairs of spines on lower surface of fourth segment; antenna 1 with tiny curved spine near base of first segment. Uropod 1 peduncle with three or four spines on inner margin. Up to 5.5 mm long. West and southwest coasts. Lower shore and shallow subtidal. Builds tube in *Corallina* turf, kelp holdfasts, among algae and hydroids
Corophium bonellii (Milne Edwards) (V.22)

25 Uropod 3 with branches shorter than peduncle (V.23)   26
– Uropod 3 with branches as long as or longer than peduncle   32

26 Third segment of antenna 1 less than half length of second segment (V.23)   27
– Third segment of antenna 1 more than half length of second segment   28

27 Pereopods 5 – 7 with propodus broad at the end, forming subchelate closure with dactyl. Hind edge of telson with pair of large, curved spines. Antennae half body length, with sparsely distributed bristles. Up to 8 mm long. Southwest and west coasts. Lower shore, among algae and in kelp holdfasts
Ampithoe gammaroides (Bate) (V.23)

– Propodus of pereopods 5 – 7 only slightly expanded at the end, not markedly subchelate. Telson without spines on hind edge. Eyes small. Up to 20 mm long; red to green with dark spots, often with white patches. Widespread and common. Intertidal. Builds tube among wide variety of algae, in red algal turfs, in holdfasts   *Ampithoe rubricata* (Montagu) (V.24)

28 Uropod 3 unbranched. Body markedly flattened from top to bottom. Head elongate, with very large eyes. Up to 10 mm long; brown to orange with darker patches. South and west coasts. Intertidal. Builds tube among algae, in algal turfs, hydroids and kelp holdfasts   *Erichthonius* species (V.25)
– Uropod 3 two-branched   29

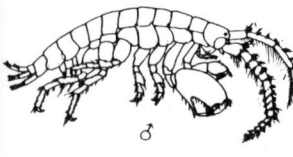

V.26

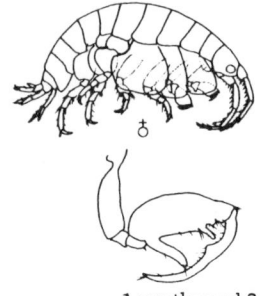

V.27

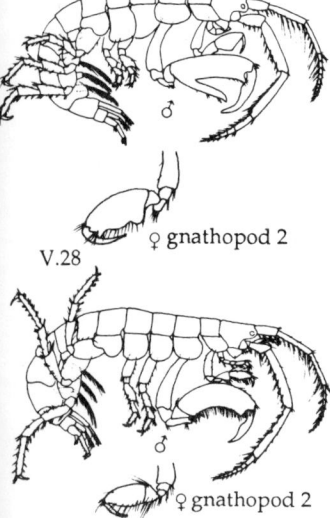

V.28

V.29

29 Accessory flagellum indistinct, reduced to minute bump. Antenna 2 very robust, with thick girdles of bristles. Up to 9 mm long; grey with brown bands. Widespread and common. Lower shore, in holdfasts, among hydroids and in algal turfs
Parajassa pelagica (Leach) (V.26)
– Accessory flagellum distinct, though short 30

30 Coxal plate 1 very small, mostly concealed by plate 2; plates 2–4 broad and elongate. Antennae with long, sparse bristles. Eyes large. Up to 3 mm long; brown. West coasts. Lower shore, in kelp holdfasts
Microjassa cumbrensis (Stebbing & Robertson) (V.27)
– Coxal plate 1 distinct; plates 1–4 of similar size, plate 6 much smaller than plate 5 31

31 Gnathopod 2 of male with single large 'thumb' on palm of propodus. Body slender, rather flattened. Antennae stout, with dense bristles. Eyes small. Up to 12 mm long; white with characteristic red-brown spots, blotches and streaks. Widespread and common. Builds tube in algal turfs, among hydroids and in holdfasts *Jassa falcata* (Montagu) (V.28)
– Gnathopod 2 of male with concave, densely hairy palm to propodus, lacking a 'thumb'. Eyes relatively large. Up to 10 mm long, sometimes larger; light yellow-green with dark spots, or white with brown bands. Widespread and common. Lower shore, building tubes among algae and in holdfasts
Ischyrocerus anguipes Kroyer (V.29)
(Females of *Jassa* and *Ischyrocerus* are hard to distinguish.)

32 Gnathopod 1 larger than gnathopod 2 (conspicuous in male, less so in female) 33
– Gnathopod 1 smaller than gnathopod 2 in both sexes 35

33 Gnathopod 1 of male with simple carpus; dactyl opposed by a short tooth on the lower margin of the propodus. Both gnathopods of male with long, dense bristles on upper surface. Up to 6 mm long; white with brown bands. Widespread and common. Intertidal and shallow sublittoral, among algae and in *Laminaria* holdfasts *Lembos websteri* Bate† (V.30)
– Gnathopod 1 of male with small propodus; dactyl opposed by one or more projecting teeth on lower margin of carpus (V.31)
34

†Females of these three species cannot be identified easily.

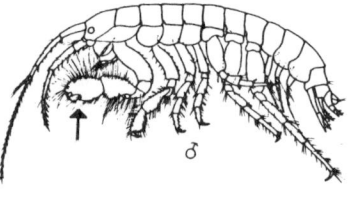

V.30

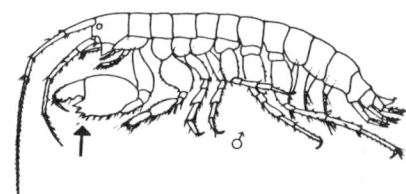

V.31

7 Identification

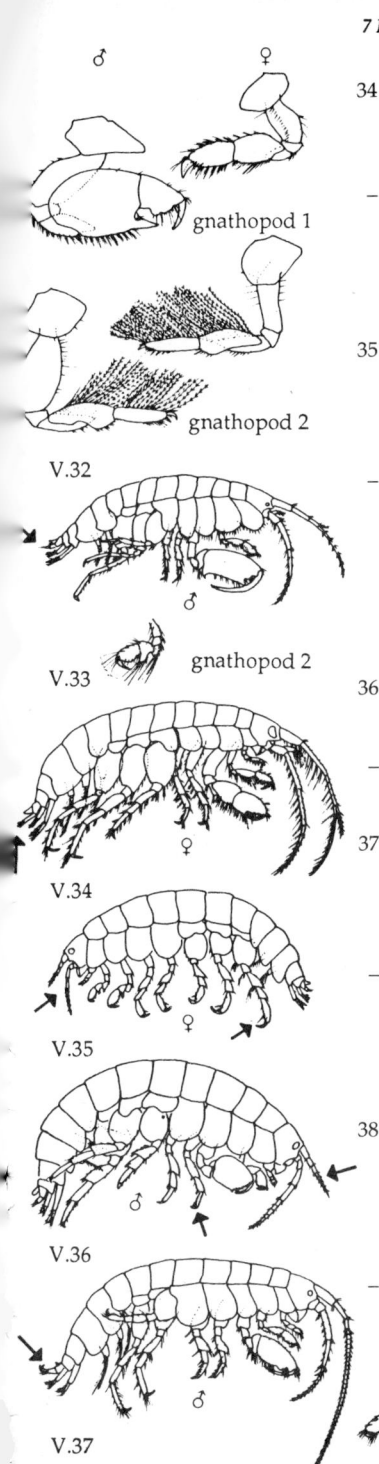

V.32
V.33
V.34
V.35
V.36
V.37

34 Gnathopod 1 of male with two distinct teeth on lower margin of carpus; gnathopod 2 of male without comb of bristles. Up to 10 mm long. Widespread and common. Intertidal, among algae and in holdfasts *Microdeutopus gryllotalpa*† Costa (V.31)

– Gnathopod 1 of male with a single large tooth on lower margin of carpus; gnathopod 2 of male with comb of coarse bristles on upper surface. Up to 8 mm long. West coasts. Lower shore, in algal turfs and *Laminaria* holdfasts
Microdeutopus versiculatus† (Bate) (V.32)

35 Uropod 3 unbranched. Coxal plates large, with fine hairs along the margins. Antennae only one-third body length. Eyes small and round. Up to 3 mm long; darkly pigmented. South and west coasts, in kelp holdfasts
Microprotopus maculatus Norman (V.33)

– Uropod 3 with two almost equal branches. Coxal plates moderately sized, with sparse, stout bristles on margins, but no fine hairs. Antennae up to half body length, with slender, elongate flagellum. Eyes large, oval. Up to 10 mm long; light yellow with dark bands. Widespread and common. Lower shore, in kelp holdfasts *Gammaropsis maculata* (Johnston) (V.34)

36 Uropod 3 unbranched. Antennae relatively short, less than one-quarter body length; antenna 2 distinctly longer than antenna 1. Pereopods 5–7 increasing in length successively 37

– Uropod 3 two-branched. Antennae long, up to half body length 38

37 Pereopods 3–7 with large, blunt spine on palm of propodus. Antenna 1 almost as long as antenna 2. Eyes small, oval, pale red. Up to 8 mm long; brownish green. Widespread and common. Lower shore in algal turfs and holdfasts
Hyale pontica Rathke (V.35)

– Pereopods 3–7 with small spines or bristles on palm of propodus. Antenna 1 a little longer than peduncle of antenna 2. Eyes moderately large, round, black. Up to 8 mm long; brown to green. Widespread and common. Intertidal, in algal turfs and holdfasts *Hyale nilssoni* (Rathke) (V.36)

38 Uropods short. Uropod 3 with branches much shorter than stem, and two stout hooks at tip of outer branch. Telson short and broad, with straight edge. Up to 10 mm long; yellow-green, or yellow with red patches. South and west coasts. Intertidal and shallow sublittoral. Among algae and in holdfasts
Sunamphitoe pelagica (Milne Edwards) (V.37)

– Uropods elongate. Uropod 3 with branches much longer than stem, outer branch without hooks at tip 39

†Females of these three species cannot be identified easily.

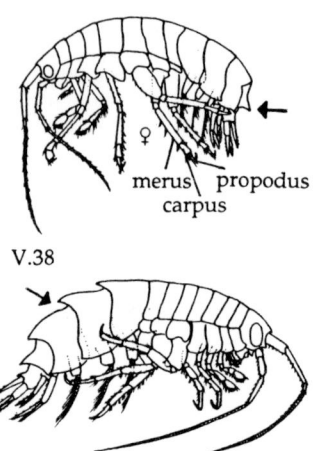

V.38

V.39

39 Pereopods 3–7 with merus longer than combined length of carpus and propodus. Body without projections on upper side, except for a small one on urosome segment 1. Up to 6 mm long; white with brown patches. Widespread and common. Lower shore; particularly associated with sponges and ascidians, often in *Laminaria* holdfasts

Tritaeta gibbosa (Bate) (V.38)

– Pereopods 3 – 7 with merus shorter than combined length of carpus and propodus. Body with well-developed projections on upper side. Up to 14 mm long; boldly coloured, red or red brown, often with white spots and black patches. Widespread and common. Lower shore, in *Laminaria* holdfasts

Dexamine spinosa (Montagu) (V.39)

VI Caprellidean amphipods

Caprellids, or skeleton shrimps, are distinctive, slow-moving amphipods, often abundant in clumps of hydroids and erect bryozoans, and in red algal turfs. Accounts of all British species are given in Chevreux & Fage (1925) and in Harrison (1944).

1 Each body segment with a pair of legs (the first two pairs constituting the gnathopods). Slender, oval gill plates at the base of the legs on segments 2, 3 and 4

Phtisica marina Slabber (VI.1)

– Segments 3 and 4 lacking legs, or with only minute rudiments. Gill plates on segments 3 and 4 only 2

2 Tiny, two-jointed legs on body segments 3 and 4. Head and first two segments with erect, forwardly directed spines

Pseudoprotella phasma (Montagu) (pl. 2.3)

– No legs on segments 3 and 4 3

3 Head bulbous, antenna 2 with tufts of short bristles on lower side *Caprella acanthifera* Leach (VI.2)

– Head more slender (VI.3, VI.4). Parallel rows of long bristles on lower side of antenna 2 4

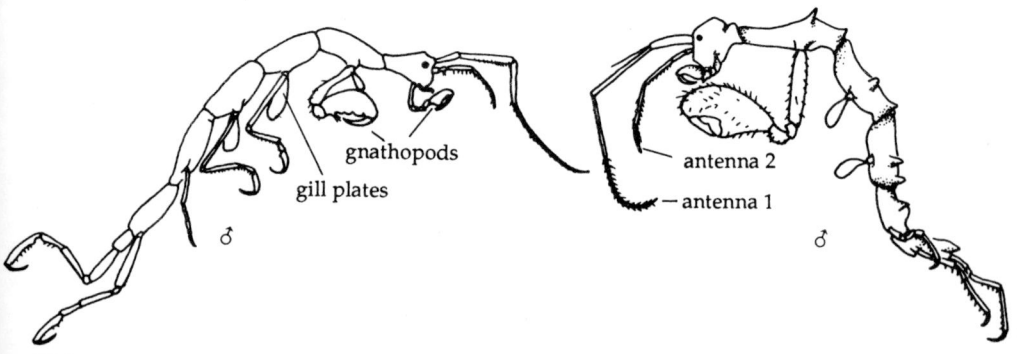

VI.1 VI.2

VI.3

VI.4

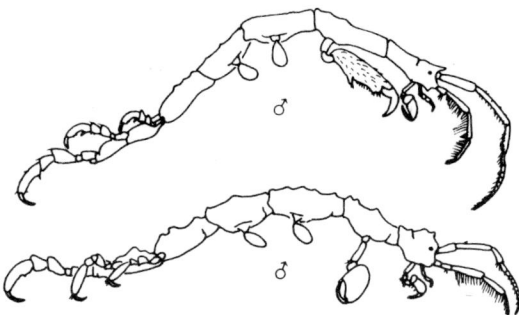

4   Head with prominent, forwardly directed process in front of eye. Gnathopod 2 arising close to hind edge of segment 2
Caprella fretensis Stebbing (VI.3)
–   Head smooth or knobbed, but without a prominent process   5

5   Head and first body segment, together, about as long as second body segment. Upper surface of head and segments 1–4 either smooth, or with a few bumps in widely spaced pairs
Caprella linearis (Linn.) (pl. 2.1, 2.2)
–   Head and first body segment, together, equivalent to about half length of second body segment. Upper surface of head and segments 1–4 typically with irregular, unpaired bumps
Caprella septentrionalis Kroyer (VI.4)

VII Pycnogonids

The few sea spiders keyed out here are those which are presently known to occur commonly in red algal turfs or as micropredators of sessile epiphytes. All British species may be identified using King (1974), and the recent AIDGAP key published by the Field Studies Council (King, 1986).

1   Chelifores and palps present (VII.3)   2
–   Chelifores or palps, or both, lacking   6

2   Palps five-jointed. Chelifores longer than proboscis   3
–   Palps with eight or nine joints. Chelifores shorter than proboscis   4

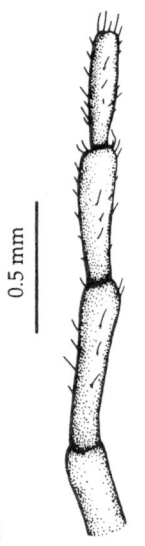

VII.1

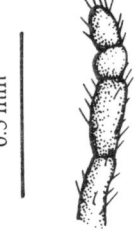

VII.2

3   Last two segments of palps of equivalent length (VII.1). Body smooth, with relatively elongate proboscis; up to 8 mm long. Legs up to 30 mm long, smooth, with a few bristles on the last segment. A carnivore, known to feed on the hydroid *Dynamena pumila* and the bryozoan *Bowerbankia*   *Nymphon gracile* Leach
–   Last segment of palp about twice length of next-to-last segment (VII.2). Body up to 5 mm long, proboscis relatively short and stout. Legs up to 20 mm long, with bristles on most segments. Lower shore, on hydroids and sponges, among a variety of algae   *Nymphon brevirostre* Hodge (VII.3)

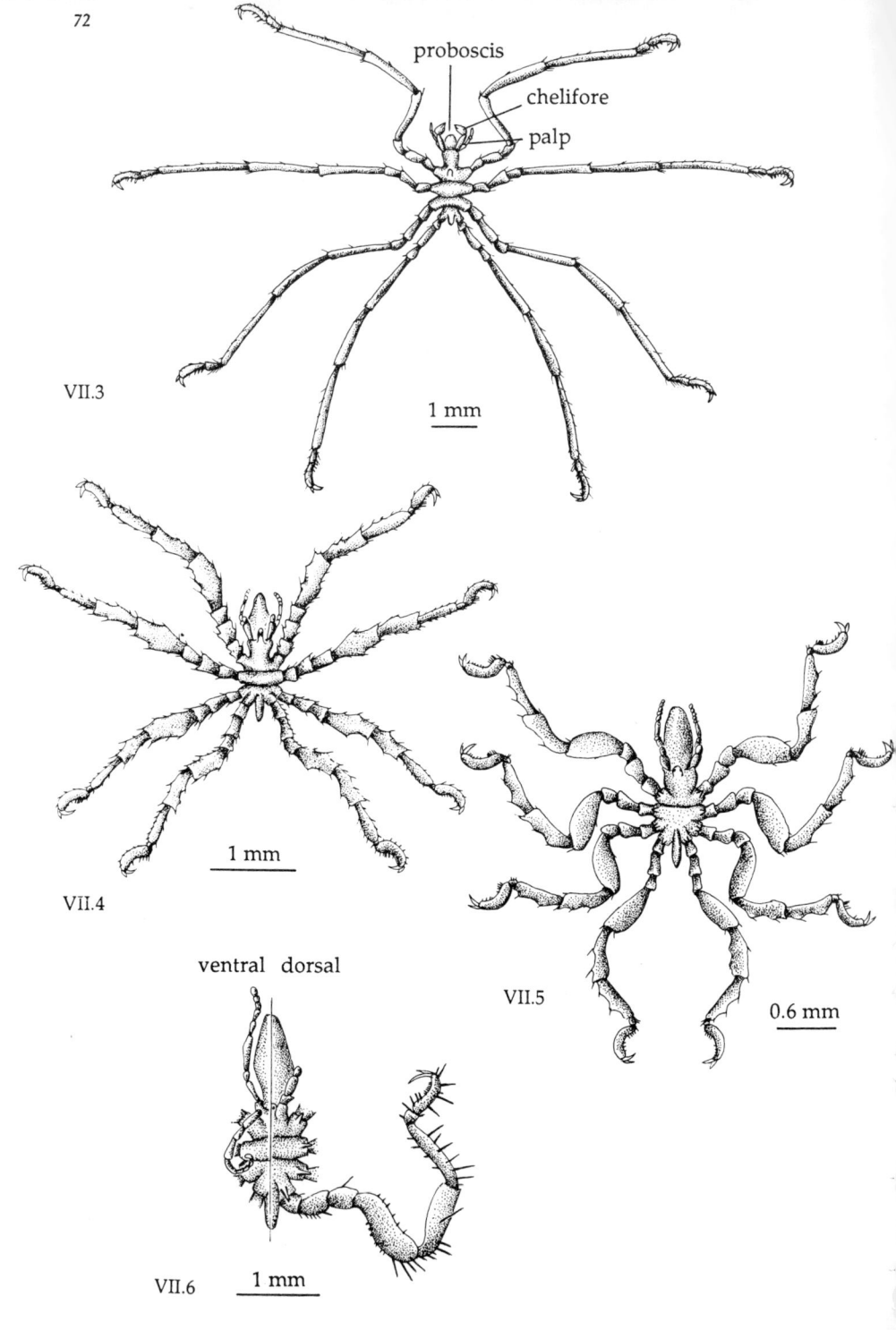

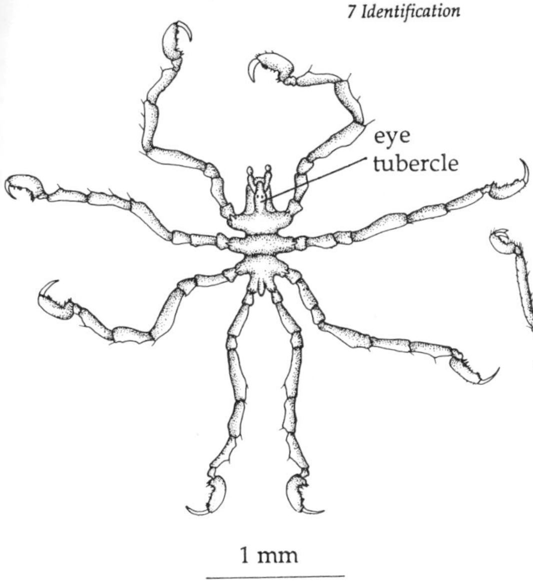

VII.7 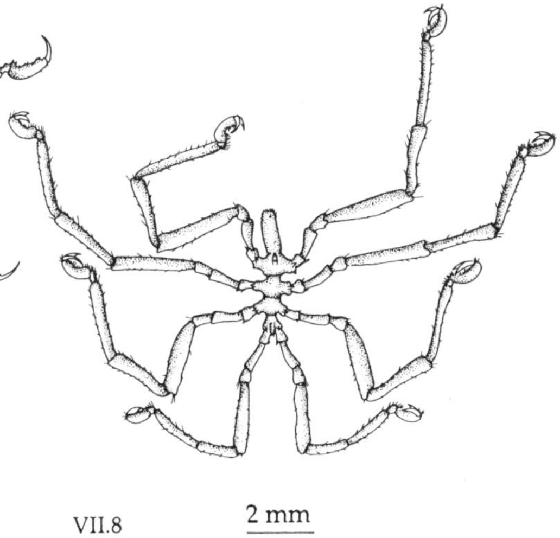 VII.8

4   Palps with eight segments. Proboscis tapered. Body smooth, up to 2 mm long; legs with spines and bumps, up to 6 mm long. On *Dynamena pumila*, and perhaps epiphytic bryozoans
    Achelia echinata Hodge (VII.4)
-   Palps with nine segments   5

5   Body with only one groove across it. Chelifores half length of proboscis. Body smooth, 2 mm long; legs with few scattered bristles, 6 mm long. On *Chondrus*, and other red algae, encrusted with the bryozoans *Flustrellidra hispida* and *Alcyonidium hirsutum*   *Achelia longipes* Hodge (VII.5)
-   Body with two grooves. Chelifores more than half length of proboscis. Body smooth, 2 mm long; legs with numerous spines and bristles, 6 mm long. Among *Chondrus* and other red algae, encrusted with *F. hispida* and *A. hirsutum*
    Achelia hispida Hodge (VII.6)

6   Chelifores present, palps lacking. Proboscis short, cylindrical; head short and wide with prominent eye tubercles. Body about 1 mm long, legs up to 3 mm. On hydroids and the bryozoan *Bowerbankia*   *Anoplodactylus* (several similar species) (VII.7)
-   Both chelifores and palps lacking. Proboscis elongate. Body slender, 2–3 mm long; legs with scattered spines and bristles 8–15 mm long. Among hydroids
    Endeis (several similar species) (VII.8)

VIII Lamellibranchs

The lamellibranchs are the largest group within the molluscan class Bivalvia. Identification often depends on opening the shells and examining their inner surfaces. Keys, descriptions and figures of all British bivalve molluscs are provided by Tebble (1976).

1 Shell thin and brittle, up to 60 mm diameter but commonly much smaller. Right valve flat with a large hole close to the umbo, left valve convex (VIII.1). Living tightly clamped to the substratum, the convex left valve uppermost 2
– Not as described. Shell attached by a horny thread, the byssus, or free, but not with one valve clamped to the substratum 3

2 Left valve with three closely spaced scars on the inner surface. Outer surface with coarse concentric sculpture, often raised as distinct scales. On kelp holdfasts
Anomia ephippium (Linn.) (VIII.1)
– Left valve with two scars close together on the inner surface, forming a single elongate shape. Outer surface with fine concentric lines. Up to 15 mm diameter. On holdfasts of kelps and other large algae *Heteranomia squamula* (Linn.) (VIII.2)

3 Shell elongate oval with radiating ribs; sharply pointed ears projecting on each side of the umbo. Shell gaping along margins revealing a fringe of red or orange tentacles. In kelp holdfasts, constructing a 'nest' of gravel and shell debris. South and west coasts only *Lima hians* (Gmelin) (VIII.3)
– Not as described 4

4 Shell with umbones at or near one end; rounded-triangular, oval or bean shaped, and inner edges without hinge teeth 5
– Umbones not at or near end, except in *Hiatella arctica*, which has hinge teeth 10

5 Shell smooth, with or without concentric or radiating lines, but without raised ribs 6
– Shell with 2 series of radiating ribs, separated by a smooth area, extending from the umbones to the lower margin 8

6 Umbones at the extreme (anterior) end of the shell. Colour blue-black. On kelp holdfasts and sometimes on other large lower shore algae, attached by byssal threads
Mytilus edulis (Linn.) (VIII.4)
– Umbones not quite at end of the shell, with the shell margin extending beyond the beaks 7

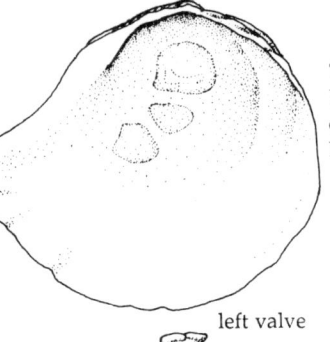

left valve

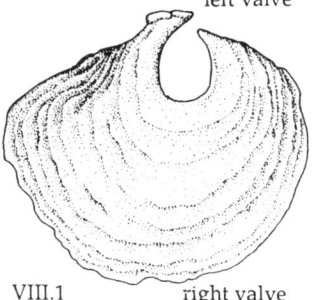

VIII.1   right valve

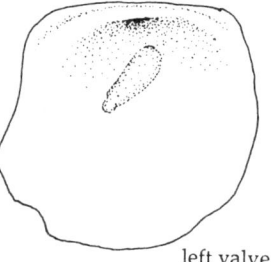

left valve

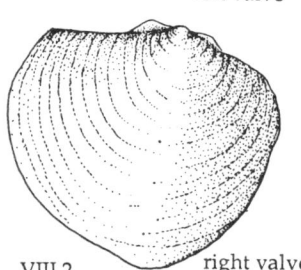

VIII.2   right valve

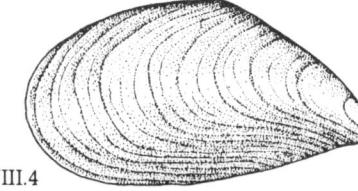

VIII.3   VIII.4

7 Identification

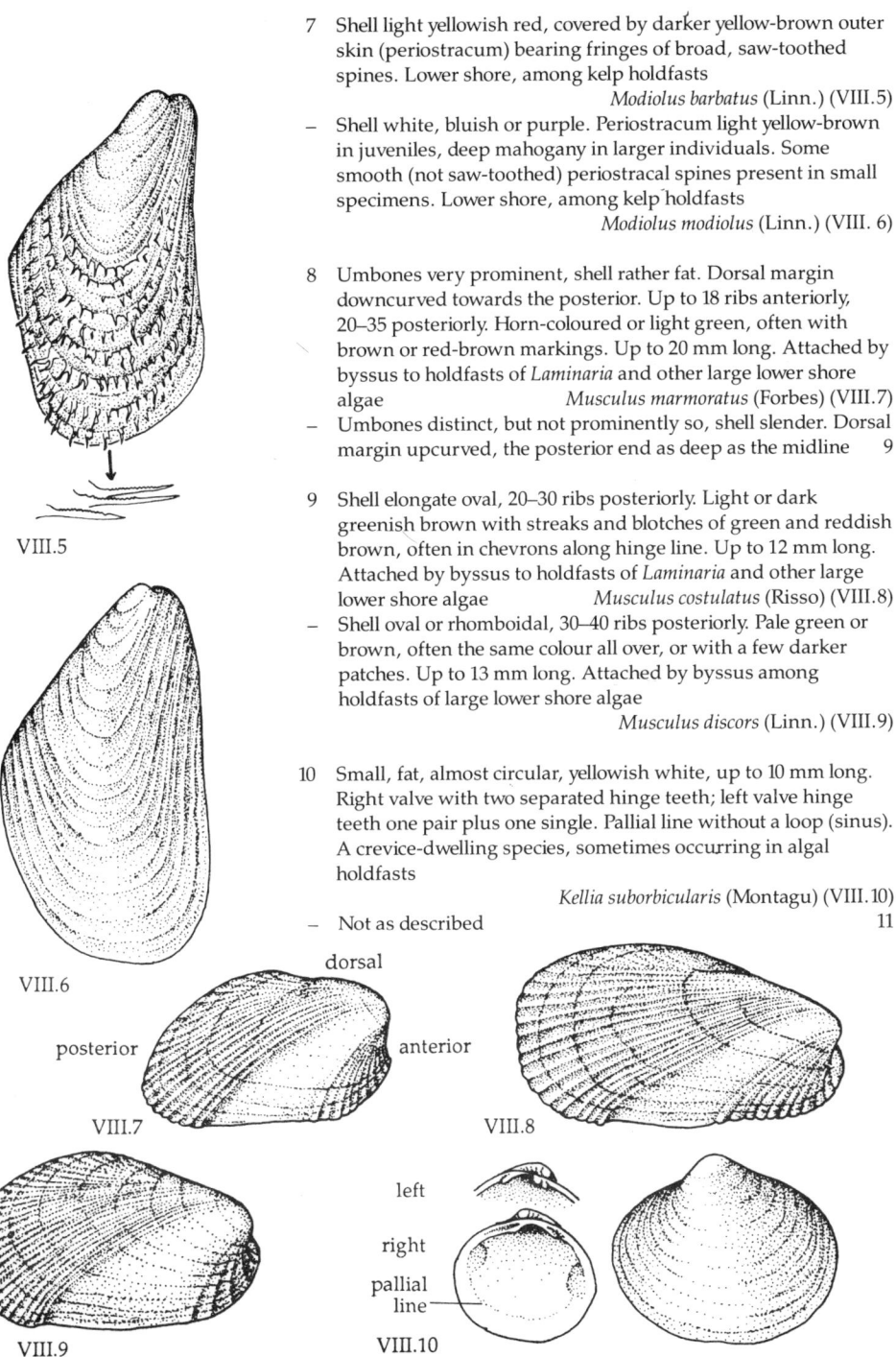

7 Shell light yellowish red, covered by darker yellow-brown outer skin (periostracum) bearing fringes of broad, saw-toothed spines. Lower shore, among kelp holdfasts
Modiolus barbatus (Linn.) (VIII.5)
– Shell white, bluish or purple. Periostracum light yellow-brown in juveniles, deep mahogany in larger individuals. Some smooth (not saw-toothed) periostracal spines present in small specimens. Lower shore, among kelp holdfasts
Modiolus modiolus (Linn.) (VIII. 6)

8 Umbones very prominent, shell rather fat. Dorsal margin downcurved towards the posterior. Up to 18 ribs anteriorly, 20–35 posteriorly. Horn-coloured or light green, often with brown or red-brown markings. Up to 20 mm long. Attached by byssus to holdfasts of *Laminaria* and other large lower shore algae   *Musculus marmoratus* (Forbes) (VIII.7)
– Umbones distinct, but not prominently so, shell slender. Dorsal margin upcurved, the posterior end as deep as the midline   9

9 Shell elongate oval, 20–30 ribs posteriorly. Light or dark greenish brown with streaks and blotches of green and reddish brown, often in chevrons along hinge line. Up to 12 mm long. Attached by byssus to holdfasts of *Laminaria* and other large lower shore algae   *Musculus costulatus* (Risso) (VIII.8)
– Shell oval or rhomboidal, 30–40 ribs posteriorly. Pale green or brown, often the same colour all over, or with a few darker patches. Up to 13 mm long. Attached by byssus among holdfasts of large lower shore algae
Musculus discors (Linn.) (VIII.9)

10 Small, fat, almost circular, yellowish white, up to 10 mm long. Right valve with two separated hinge teeth; left valve hinge teeth one pair plus one single. Pallial line without a loop (sinus). A crevice-dwelling species, sometimes occurring in algal holdfasts
Kellia suborbicularis (Montagu) (VIII.10)
– Not as described   11

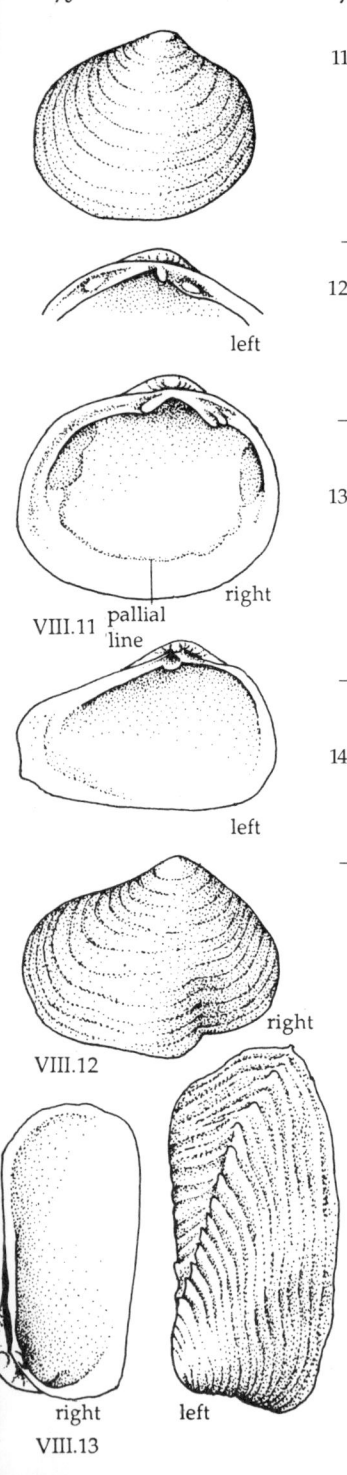

11 Fat, oval, yellowish brown, tinged with pink or red, darkest at the umbones, up to 3 mm long. Right valve with single anterior and posterior lateral teeth; left valve with one small hinge tooth below the umbo and single anterior and posterior lateral teeth. Pallial line without a sinus. Intertidal, attached by byssus, in crevices and, often in large numbers, in *Lichina pygmaea* and upper shore fucoid algae *Lasaea rubra* (Montagu) (VIII.11)
– Not as described 12

12 Hinge ligament internal and external, the internal portion accommodated within a conspicuous pit (chondrophore). No hinge teeth. Shell irregularly oval, up to 20 mm long. Attached by byssus in algal holdfasts on lower shore
 Sphenia binghami Turton (VIII.12)
– Hinge ligament external only, no chondrophore. Hinge teeth present 13

13 Shell solid, elongate, roughly rectangular; umbones close to one end, the other end squared. Right valve with one small tooth, left valve with two: all indistinct and liable to wear. Sculpture of coarse concentric ridges; dull white with yellow-brown periostracum. A crevice-dwelling species, attached by byssus, frequently in holdfasts of large, lower shore algae
 Hiatella arctica (Linn.) (VIII.13)
– Shell oblong or oval, umbones nearer middle of shell. Each valve with three hinge teeth. Pallial line with a sinus 14

14 Shell oval, smooth, plump, not exceeding 3 mm long. Dull white to light brown. Intertidal only, attached by byssus, in crevices, often abundant in algal holdfasts
 Turtonia minuta (Fabricius) (VIII.14)
– Shell elongate, oblong, with conspicuous concentric ridges, up to 25 mm long. Yellow-white to light brown. In crevices, and kelp holdfasts, on the lower shore *Notirus irus* (Linn.) (VIII.15)

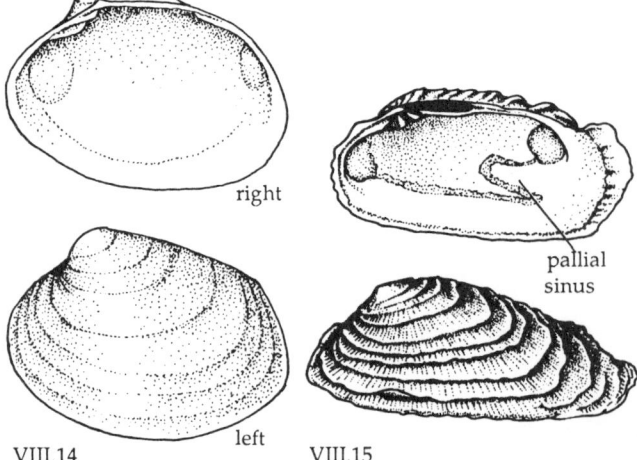

IX Prosobranchs

The prosobranch gastropods include the winkles, limpets and all sea snails. All British species may be identified using Graham (1971). A comprehensive account of prosobranch biology has been provided by Fretter & Graham (1962). In some cases it is important to be able to see the living animal, as identification may depend on the presence or absence of tentacles around the foot, or on the edges of the mantle – a fold of tissue seen projecting from the shell above the crawling animal.

1 Shell conical, without a coiled spire 2
– Shell with spire of two whorls or more, elongate and spindle-shaped, or squat, with a depressed spire on one side of the shell; or oval and domed, the spire partly or wholly hidden by the body whorl, with a slit-like aperture on the lower side (cowries) 3

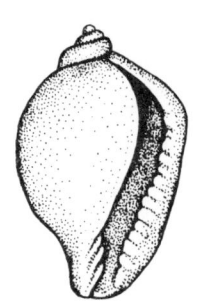

1 cm
IX.1

2 Shell smooth, translucent, with conspicuous blue rays
   Up to 6mm long, blue rays well marked; on frond and stipe of *Laminaria*   *Patina pellucida* (Linn.) (pl. 6.4)
   Up to 10 mm long, blue rays faintly showing at apex of shell; within holdfast of *Laminaria*   *Patina pellucida* variety *laevis*
– Shell thicker, opaque, without blue rays; with radial ribs, and radiating bands of colour: brown, green, grey, yellow; opaque
   Patella species (fig. 3, p. 2)
(Juvenile rock limpets often occur on haptera of large kelp holdfasts.)

3 Shell consisting largely of an inflated body whorl. Aperture long and narrow, extending whole length of body whorl, outer lip thickened and angular. Mantle lobes extending to sides, partly enveloping body whorl. Up to 10 mm long. Feeds on compound ascidians *Botryllus* and *Botrylloides*. South and west coasts   *Erato voluta* (Montagu) (IX.1)
– Not as described 4

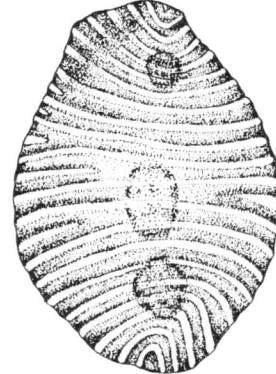

5 mm
IX.2

4 Cowries: shell strongly convex, ribbed, with slit-like aperture extending whole length of flat side. Up to 10 mm long. On *Botryllus*, *Botrylloides* and other compound ascidians, depositing egg cases on food organisms. Often on large kelp holdfasts
   Pinkish white, with three black spots on convex side
   Trivia monacha (da Costa) (IX.2)
   Pinkish white, no spots   *Trivia arctica* (Montagu) (IX.3)
– Not as described 5

5 Shell with a well-marked groove (siphonal canal) in the lip of the aperture; up to 7 mm long, cylindrical, abruptly tapered at tip of spire. Chestnut brown. West and southwest coasts. On sponges; in *Corallina* and red algal turfs
   Cerithiopsis tubercularis (Montagu) (pl. 7.3)
– Shell without a siphonal canal 6

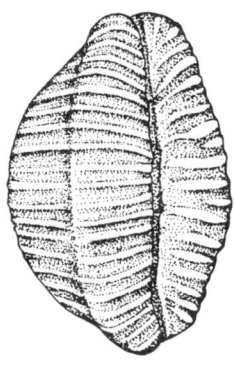

5 mm
IX.3

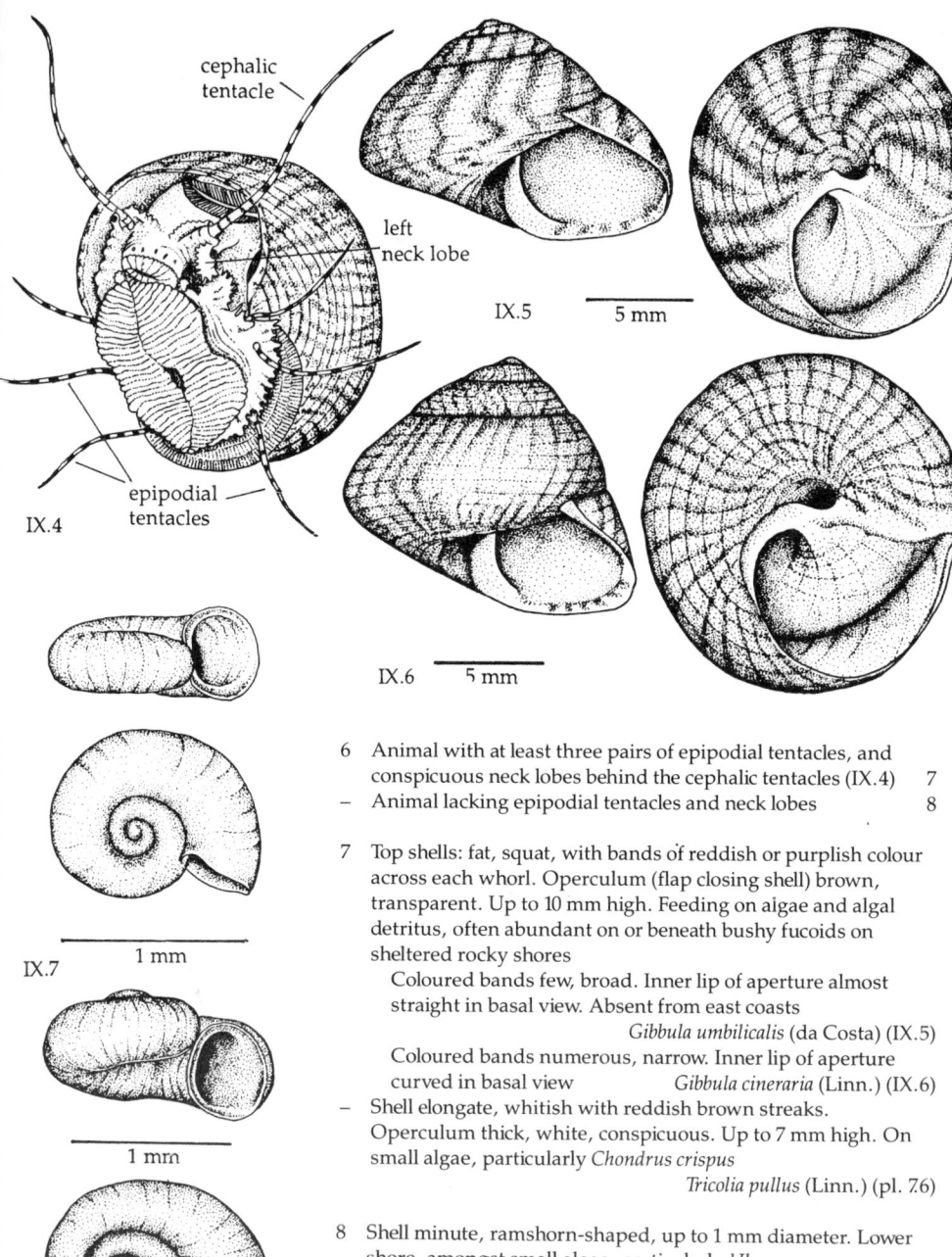

6 Animal with at least three pairs of epipodial tentacles, and conspicuous neck lobes behind the cephalic tentacles (IX.4)   7
– Animal lacking epipodial tentacles and neck lobes   8

7 Top shells: fat, squat, with bands of reddish or purplish colour across each whorl. Operculum (flap closing shell) brown, transparent. Up to 10 mm high. Feeding on algae and algal detritus, often abundant on or beneath bushy fucoids on sheltered rocky shores

     Coloured bands few, broad. Inner lip of aperture almost straight in basal view. Absent from east coasts
                             Gibbula umbilicalis (da Costa) (IX.5)
     Coloured bands numerous, narrow. Inner lip of aperture curved in basal view   *Gibbula cineraria* (Linn.) (IX.6)
– Shell elongate, whitish with reddish brown streaks. Operculum thick, white, conspicuous. Up to 7 mm high. On small algae, particularly *Chondrus crispus*
                             Tricolia pullus (Linn.) (pl. 7.6)

8 Shell minute, ramshorn-shaped, up to 1 mm diameter. Lower shore, amongst small algae, particularly *Ulva*
     Shell flat and disc-like   *Omalogyra atomus* (Philippi) (IX.7)
     Shell with a low spire   *Skeneopsis planorbis* (Fabricius) (IX.8)
– Not as described   9

7 Identification

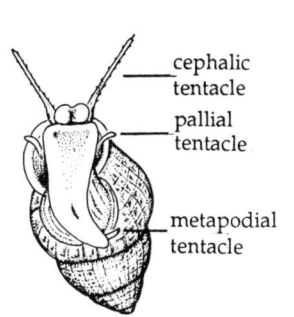

IX.9

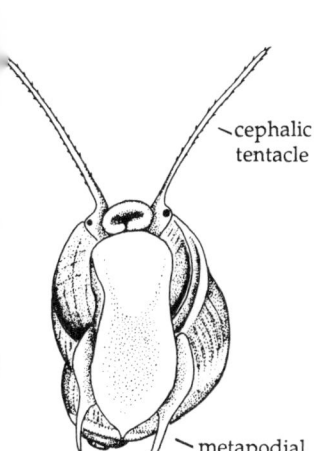

IX.10

9	Animal with single or paired metapodial tentacle (IX.9, IX.10)	10
–	Animal without metapodial tentacles	18

10	A single metapodial tentacle, visible at hind end of animal when crawling (IX.9)	11
–	Paired metapodial tentacles, pointing backwards (IX.10)	16

11 Paired pallial tentacles present, situated just behind head on each side (IX.9). Shell with spiral and longitudinal ridges, giving a net-like effect. Up to 3 mm long. Among epiphytic bryozoans on kelp holdfasts and stipes
Alvania punctura (Montagu) (IX.11)
– Pallial tentacle present on right side only, or absent. Shell smooth, ribbed or grooved, but without net-like pattern 12

12 No pallial tentacle. Shell with ribs and fine grooves on body whorl only; tip of spire orange; inside edge of outer lip of aperture violet. Up to 4 mm long. Among small red algae on lower shore *Rissoa lilacina* Recluz (IX.12)
– A single pallial tentacle on right side of animal 13

13 Shell with fine spiral lines. Ribs, if present, developed only on the upper edge of each whorl. Outer lip of aperture without a thickening rib. Up to 3 mm long. Among small algae, bryozoans and hydroids on lower shore
Onoba semicostata (Montagu) (IX.13)
– Not as described 14

14 Shell smooth or ribbed, with a conspicuous dark comma-shaped mark on upper part of body whorl, close to outer lip. Up to 4 mm long. Often abundant on filamentous red algae. Also in kelp holdfasts and among bryozoan and hydroid clumps *Rissoa parva* (da Costa) (IX.14, pl. 72)
– Not as described 15

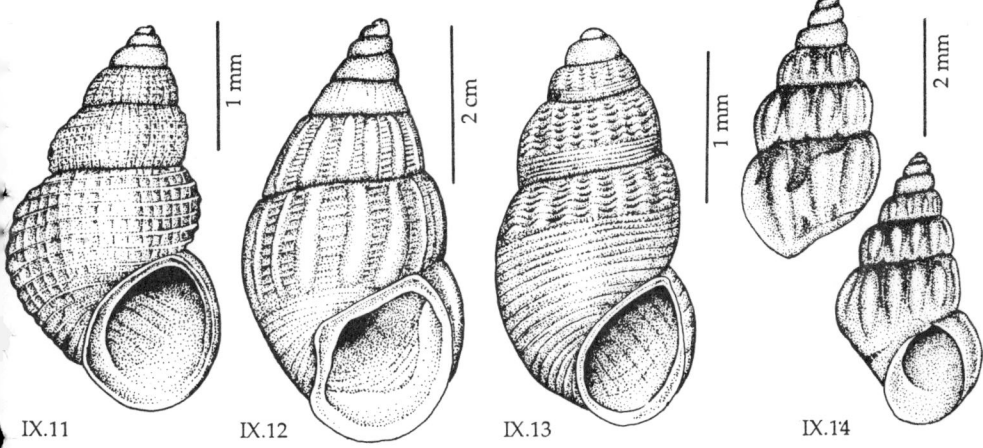

IX.11 IX.12 IX.13 IX.14

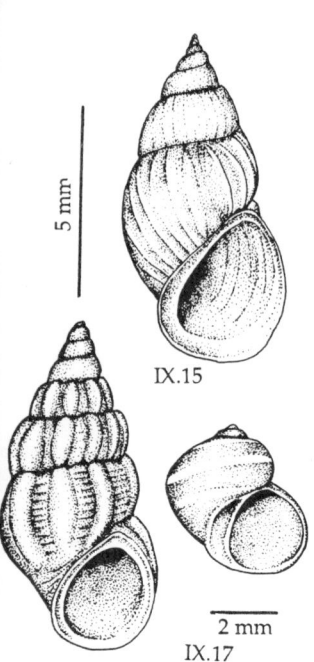

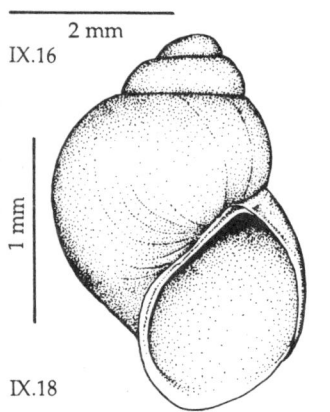

15 Aperture distinctly flared. Body whorl comprising two-thirds height; faint ribs present on body whorl only. Up to 7 mm long. On small seaweeds, lower shore
Rissoa membranacea (Adams) (IX.15)
– Aperture not flared. Body whorl comprising about half total height; well-developed ribs on all whorls. Up to 4 mm long. On *Codium* and other small algae. South and west coasts only
Rissoa guerini (Recluz) (IX.16)

16 Body whorl expanding broadly, aperture height almost equivalent to total shell height. Up to 8 mm long. On lower shore fucoids and kelps   *Lacuna pallidula* (da Costa) (pl. 7.4)
– Aperture height equivalent to half, or less, shell height   17

17 Shell conical, with six whorls, aperture markedly angular. Up to 6 mm long. On fucoid algae   *Lacuna vincta* (Montagu) (pl. 7.1)
– Shell globular, with three or four whorls, aperture rounded. Up to 5 mm long. On *Chondrus crispus* and other small algae. South and west coasts   *Lacuna parva* (da Costa) (IX.17)

18 Shell tiny (1–2 mm long), transparent; rather globular, with a marked umbilicus. Cephalic tentacles forked. On small red and green algae in lower shore pools   *Rissoella* species (IX.18)
– Not as described   19

19 Shell minute (1 mm long), pale-coloured, with spiral brown bands. On red and green algae in rock pools. South and west coasts   *Cingulopsis fulgida* (Adams) (IX.19)
– Not as described   20

20 Shell elongate, dark red, or with broad, dark red bands. Up to 3 mm long. Operculum deep crimson, with concentric lines. Lower shore, in red algal turfs. South and west coasts
Barleeia unifasciata (Montagu) (pl. 7.5)
– Shell squat, exceeding 3 mm long. Operculum horny, with spiral lines   21

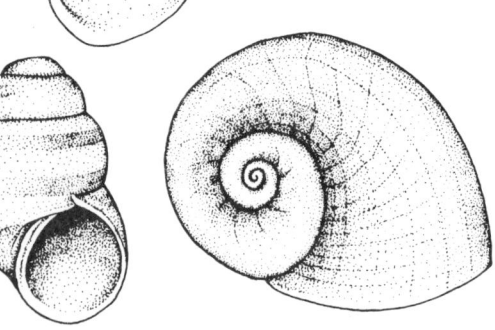

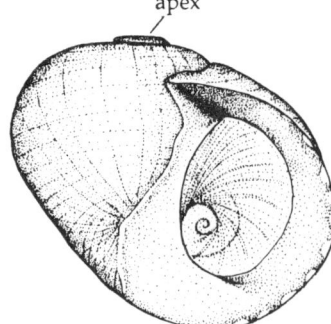

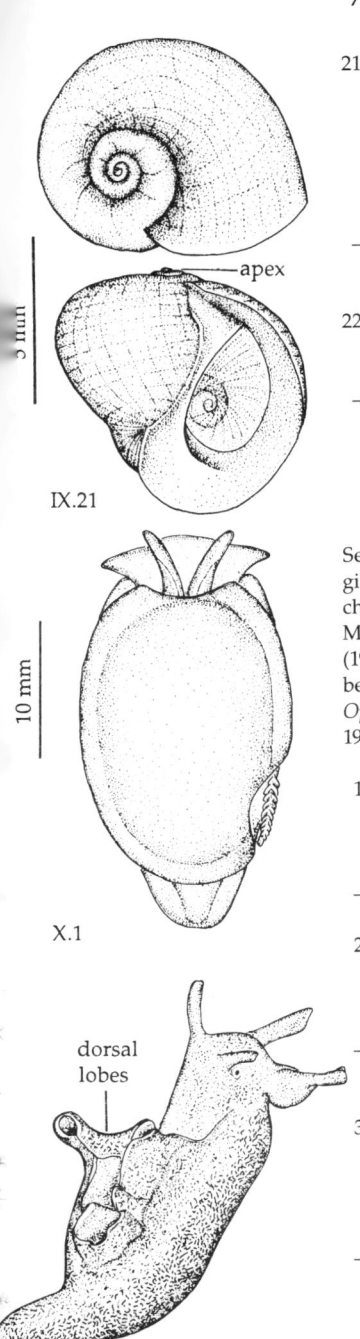

21 Shell with pointed spire, ornamented with spiral lines and grooves. Aperture flared. Colour variable. Up to 8 mm long. Middle and upper shore, among *Pelvetia canaliculata* and *Fucus* species. All coasts, common
Littorina saxatilis (Olivi) (fig. 6, p. 5)
(*L. arcana* Hannaford-Ellis is a similar species which occurs high among barnacles and lichens on exposed shores.)

— Shell globose, spire depressed; aperture as wide as shell height   22

22 Aperture relatively broad, outer lip arising well below level of apex. Shell with obliquely oval outline in apertural view
Littorina obtusata (Linn.) (IX.20)

— Aperture relatively narrow, outer lip arising close to apex. Shell with obliquely drop-shaped outline in apertural view
Littorina mariae Secchi & Rastalli (IX.21)

X Opisthobranchs

Sea slugs are best identified alive, as colours, and structures such as gills which are only everted in the living animal, are important characters. Fig. 47 shows the main features of sea slug morphology. Most British sea slugs may be identified using Thompson & Brown (1976) and good accounts of the biology and ecology of these often beautiful animals may be found in the two volumes of *Biology of Opisthobranch Molluscs* (Thompson, 1976, and Thompson & Brown, 1984).

1 Soft, flat and rather formless, with an internal shell hard to the touch. Pale yellow to orange, up to 60 mm long. A common predator of compound ascidians
Berthella plumula (Montagu) (X.1)

— Not as described   2

2 Body with prominent, paired, dorsal lobes. Elongate, slender; red, dark green, brown or purplish black, up to 30 cm long. Small individuals occur intertidally feeding on various seaweeds   *Aplysia punctata* (Cuvier) (X.2)

— Body with or without dorsal processes, but lacking paired lobes   3

3 Smooth, elongate; head rather square, lacking tentacles. Conspicuous gills at hind end on right side. Dark brown with pale patches on head and tail; up to 6 mm long. In rock pools, feeding on small algae, particularly *Codium*
Runcina coronata (Quatrefages) (X.3)

— Not as described   4

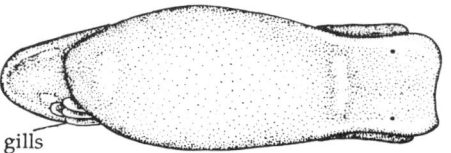

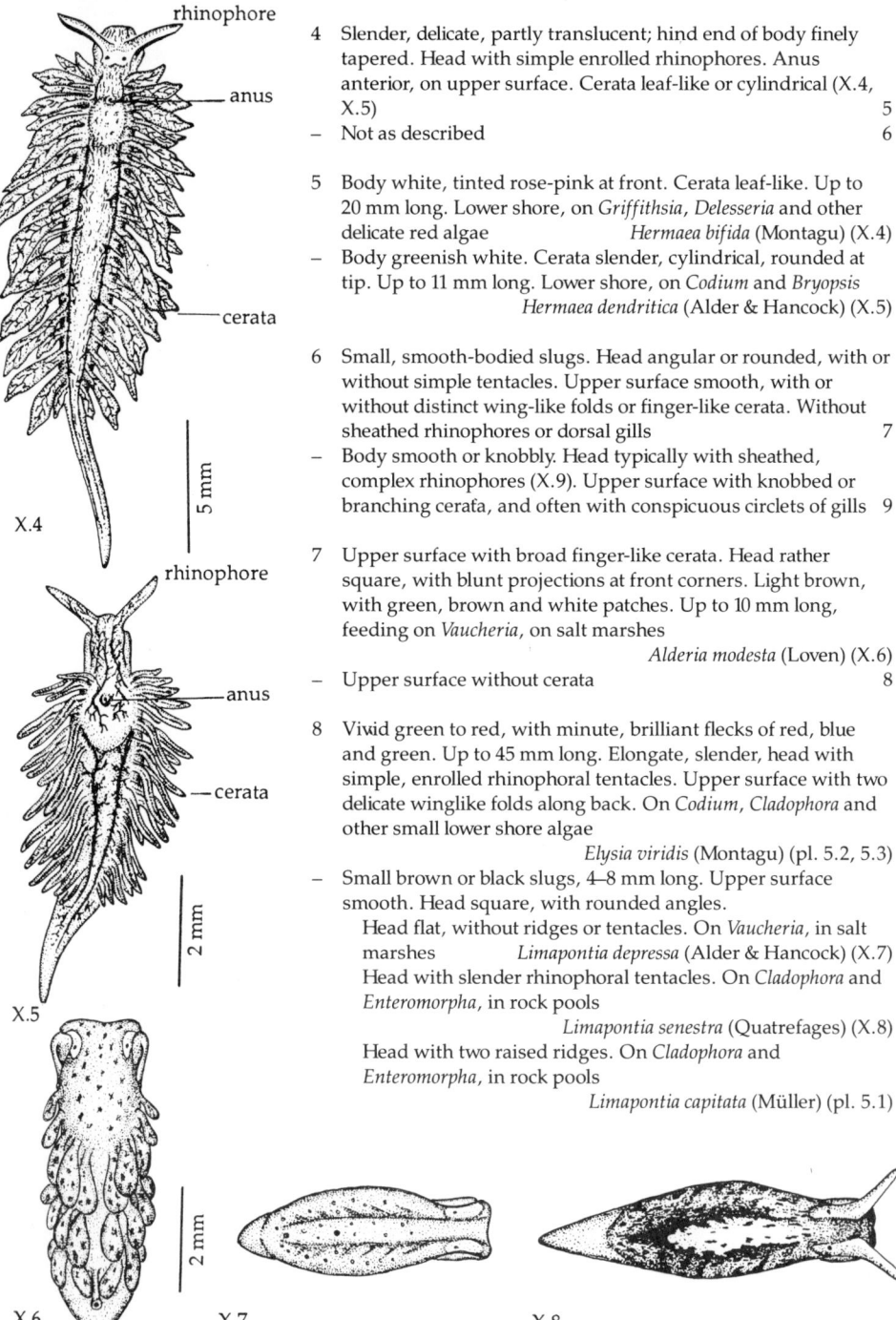

4   Slender, delicate, partly translucent; hind end of body finely tapered. Head with simple enrolled rhinophores. Anus anterior, on upper surface. Cerata leaf-like or cylindrical (X.4, X.5) 5
–   Not as described 6

5   Body white, tinted rose-pink at front. Cerata leaf-like. Up to 20 mm long. Lower shore, on *Griffithsia*, *Delesseria* and other delicate red algae   *Hermaea bifida* (Montagu) (X.4)
–   Body greenish white. Cerata slender, cylindrical, rounded at tip. Up to 11 mm long. Lower shore, on *Codium* and *Bryopsis*
    Hermaea dendritica (Alder & Hancock) (X.5)

6   Small, smooth-bodied slugs. Head angular or rounded, with or without simple tentacles. Upper surface smooth, with or without distinct wing-like folds or finger-like cerata. Without sheathed rhinophores or dorsal gills 7
–   Body smooth or knobbly. Head typically with sheathed, complex rhinophores (X.9). Upper surface with knobbed or branching cerata, and often with conspicuous circlets of gills 9

7   Upper surface with broad finger-like cerata. Head rather square, with blunt projections at front corners. Light brown, with green, brown and white patches. Up to 10 mm long, feeding on *Vaucheria*, on salt marshes
    Alderia modesta (Loven) (X.6)
–   Upper surface without cerata 8

8   Vivid green to red, with minute, brilliant flecks of red, blue and green. Up to 45 mm long. Elongate, slender, head with simple, enrolled rhinophoral tentacles. Upper surface with two delicate winglike folds along back. On *Codium*, *Cladophora* and other small lower shore algae
    Elysia viridis (Montagu) (pl. 5.2, 5.3)
–   Small brown or black slugs, 4–8 mm long. Upper surface smooth. Head square, with rounded angles.
    Head flat, without ridges or tentacles. On *Vaucheria*, in salt marshes   *Limapontia depressa* (Alder & Hancock) (X.7)
    Head with slender rhinophoral tentacles. On *Cladophora* and *Enteromorpha*, in rock pools
        Limapontia senestra (Quatrefages) (X.8)
    Head with two raised ridges. On *Cladophora* and *Enteromorpha*, in rock pools
        Limapontia capitata (Müller) (pl. 5.1)

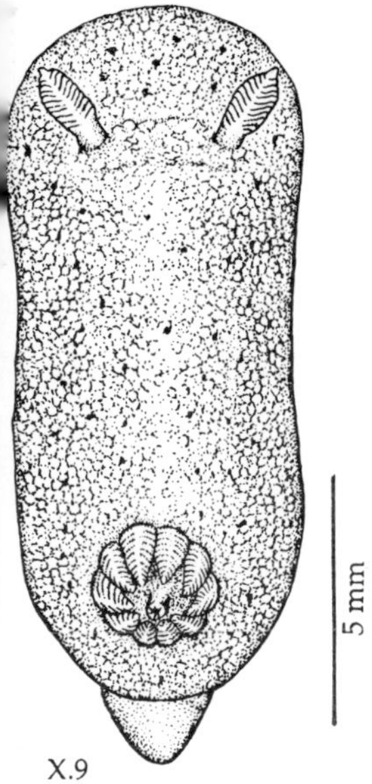

9   Gills forming a circle around the anus, at hind end of upper
    surface (X.9)   10
–   Gills, when present, not forming a circle around the anus   21

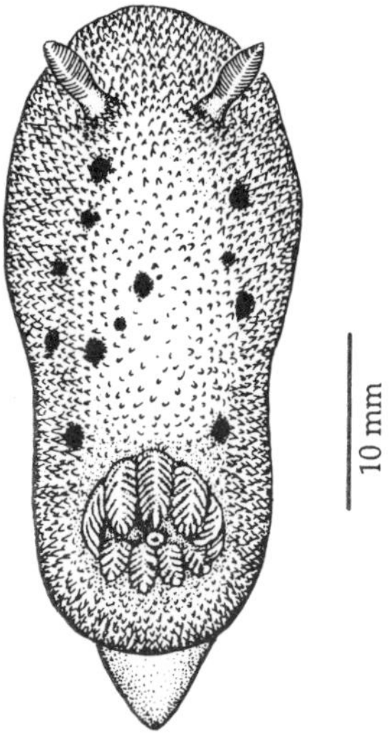

10 When disturbed, the gills retract all together into a deep
    cavity   11
–   When disturbed, the gills contract one by one   13

11 Upper surface with coarse papillae of varying sizes, with
    rough texture. Animal large, solid, fleshy. Colour variable, but
    always an irregular patchwork of yellow, brown, green, white or
    pinkish blotches. Commonly 60–80 mm long, but up to 120 mm.
    Feeds on sponges, particularly *Halichondria*, often on kelp
    holdfasts   *Archidoris pseudoargus* (Rapp) (pl. 5.4)
–   Upper surface with small, uniform papillae, having a velvety
    texture. Colour uniform, with a few dark spots   12

12 Bright red, with scattered black spots and a yellow patch
    between the rhinophores. Up to 15 mm long. On red sponges.
    West coasts only   *Rostanga rubra* (Risso) (X.9)
–   Light brown, with a few darker brown spots. Up to 55 mm
    long. On sponges, particularly *Halichondria*
                                       Jorunna tomentosa (Cuvier) (X.10)

13 Rather quadrate slugs, tapered behind. Head with prominent
    oral tentacles (X.11). A marked ridge along the back, also
    scattered, irregular papillae   14
–   Not as described   15

14 Body white, with yellow and pink patches, rhinophores yellow.
    Small, conical papillae sparingly developed on each side of
    ridge along back. Up to 27 mm long. Juveniles feeding on
    Alcyonidium and other Bryozoa, adults on *Botryllus* and other
    compound ascidians   *Goniodoris nodosa* (Montagu) (X.11)
–   Body reddish brown, with white spots, rhinophores brown.
    Coarse bumps developed over whole of body. Up to 38 mm
    long. Feeds on compound ascidians, particularly *Botryllus* and
    Botrylloides   *Goniodoris castanea* (Alder & Hancock) (X.12)

15 Generally oval, markedly convex slugs, rounded behind.
    Upper surface entirely covered with closely spaced, rounded or
    pointed papillae of various sizes   16
–   Elongate, tapered slugs. Front margin of mantle drawn out into
    short knobs or longer finger-like processes; hind margin with
    paired, simple or branched lobes (X.16)   20

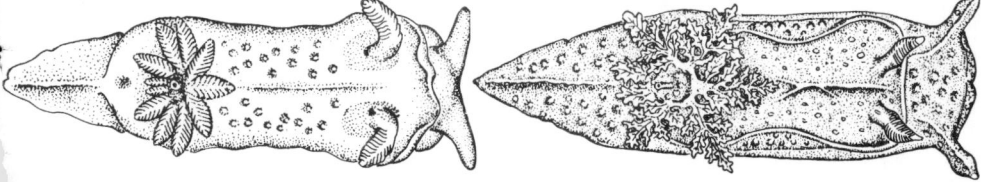

16 Rhinophores with short, frilled sheaths. Mantle papillae elongate, conical, soft. Commonly up to 30 mm long. Uniformly white, grey or brown, of varying shades. A predator of encrusting Bryozoa, particularly *Flustrellidra hispida* and *Alcyonidium* species
Acanthodoris pilosa (Muller) (X.13)

– Rhinophores without sheaths. Mantle papillae short, rounded, stiff   17

17 Upper surface with brown patches   18
– Upper surface the same colour all over, white or yellow   19

18 Mantle papillae relatively large, club-shaped. Gill circlet large, with up to 29 gills. Reaching 40 mm long. Adults feeding on acorn barnacles and encrusting, calcified bryozoans such as *Cryptosula* and *Umbonula*, juveniles on barnacles.
Onchidoris bilamellata (Linn.) (X.14)

– Mantle papillae small, spiny. Gill circlet with about 12 gills. Up to 9 mm long. A predator of encrusting bryozoans, such as *Escharoides* and *Microporella*, on kelp holdfasts
Onchidoris pusilla (Alder & Hancock) (X.15)

19 Mantle papillae elongate, tapered or club-shaped. Body yellow, pale yellow or white. Up to 17 mm long. Feeds on the bryozoan *Electra pilosa*   *Adalaria proxima* (Alder & Hancock) (pl. 5.5)

– Mantle papillae large and rounded. Body usually creamy white, rarely pale yellow. Up to 14 mm long. Feeds on the bryozoans *Electra*, *Membranipora* and *Alcyonidium*
Onchidoris muricata (Muller) (pl. 5.6)

20 Front margin of mantle with four to six smooth, finger-like processes, hind margin with two smooth, backwardly directed processes. Body white, with yellow or orange blotches and a few black streaks. Up to 39 mm long. A predator of *Membranipora*, on kelp fronds
Polycera quadrilineata (Muller) (X.16)

– Front margin of mantle with a fringe of small knobs, hind margin with paired, branching processes. Body green or yellow, with large, pale-coloured papillae. Up to 20 mm long. A predator of encrusting, calcified Bryozoa, including *Schizoporella* and *Cryptosula*   *Palio dubia* (M. Sars) (X.17)

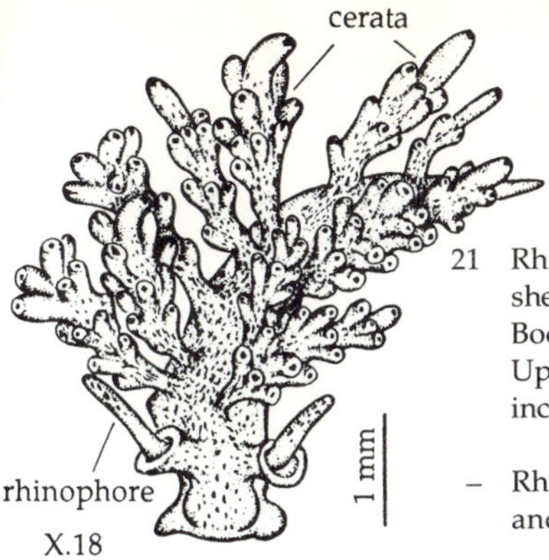

21 Rhinophores simple, finger-like, with trumpet-shaped sheaths. About eight pairs of club-shaped, knobbly cerata. Body yellow or white, with red or purple spots and blotches. Up to 12 mm long. Common on a number of hydroids, including *Sertularia*, *Obelia* and *Tubularia*
Doto coronata (Gmelin) (X.18)
– Rhinophores smooth or ridged, without sheaths; or with ridges and conspicuous basal sheaths   22

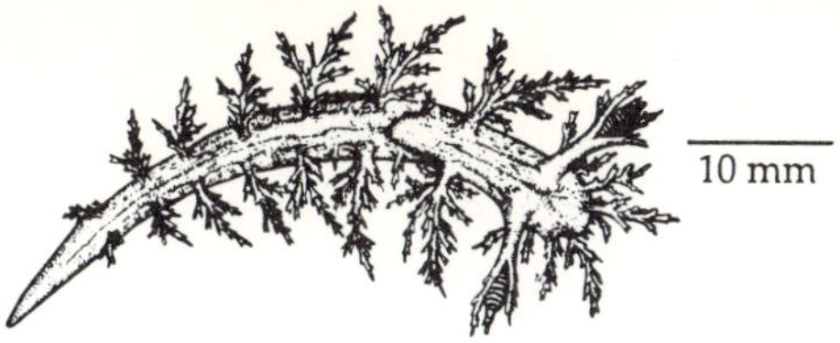

22 Rhinophores with transverse ridges, sheaths with frilled edges. Cerata and head processes richly branched. Large (up to 10 cm long); body white, with orange, red and brown blotches. A common predator of large hydroids, including *Tubularia* and *Sertularia*   *Dendronotus frondosus* (Ascanius) (X.19)
– Not as described   23

23 Rhinophores with transverse ridges. Cerata in dense clusters on each side of the middle of the back. Body white, tinged with pink. Up to 50 mm long. Feeds on a broad range of hydroids
Facelina coronata (Forbes & Goodsir) (X.20)
– Rhinophores smooth, or with faint knobs; without marked transverse ridges   24

24 Propodium with paired tentacles (X.21)   25
– With oral tentacles and rhinophores only; no propodial tentacles   26

25 Small, flattened, with inconspicuous propodial tentacles; cerata in paired clusters. Up to 13 mm long. White or cream-coloured, the cerata pale yellow to brown. A predator of fish eggs; may occur in the holdfast of *Saccorhiza* on eggs of the Clingfish *Lepadogaster*   *Calma glaucoides* (Alder & Hancock) (X.21)
– Body slender, elongate; propodial tentacles well developed, rhinophores with wrinkled or slightly pimply surface. Cerata in distinct clusters. Up to 40 mm long. Body white or violet, sometimes with white lines or blotches; cerata red or orange, often with white tips. Predators of hydroids
Coryphella species (X.22)

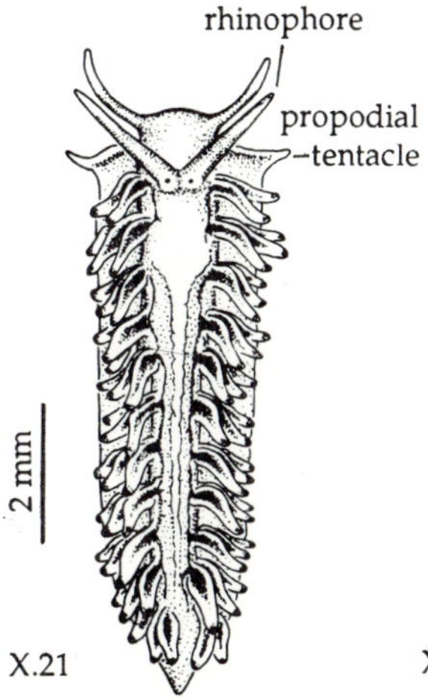

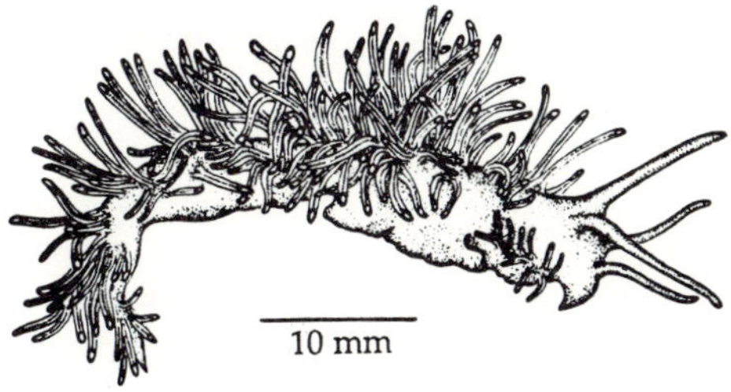

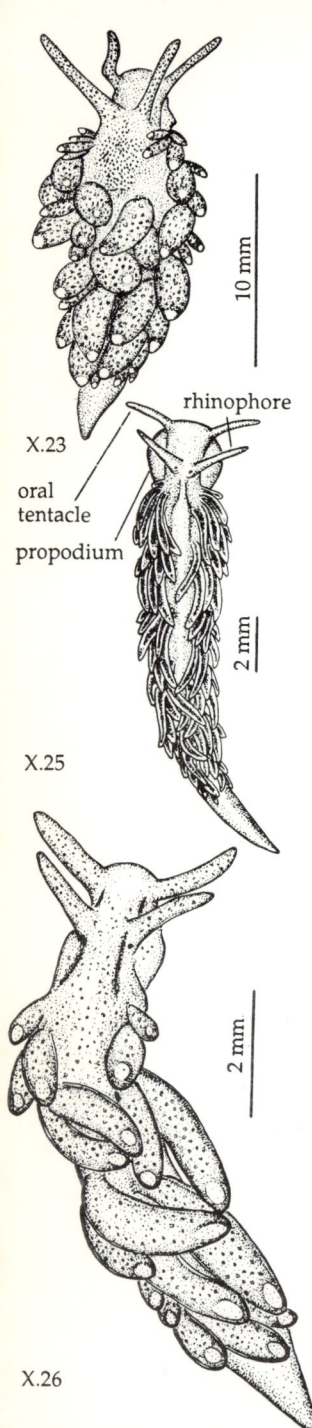
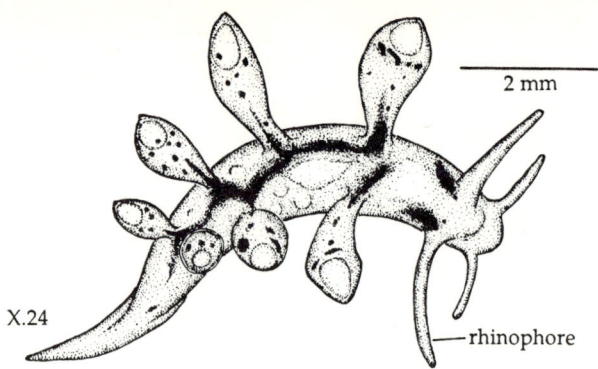

26 Cerata club-shaped, frequently banded with colour. 10–30 mm long. Feeding on hydroids among kelp holdfasts
Eubranchus species (X.23)
– Cerata stout or slender, but not club-shaped   27

27 Slender, small, with relatively long rhinophores; up to eight widely spaced cerata, the front two opposite, the rest alternating to left and right of the middle of the back. Body translucent white, cerata greenish with white tips and a red ring just below the tip. Up to 8 mm long. On *Obelia* and other small hydroids, among kelp holdfasts
Tergipes tergipes (Forskal) (X.24)
– Slender-bodied slugs, with numerous cerata in rows across the body   28

28 Propodium rounded, rhinophores and oral tentacles of similar length. Cerata reddish orange, rhinophores pink or orange. Up to 21 mm long. On *Tubularia* and other hydroids
Cuthona nana (Alder & Hancock) (X.25)
– Propodium rather angular, rhinophores and oral tentacles of unequal length. Cerata green or brown, with or without bands of blue, orange or yellow. 10–18 mm long. On a wide range of hydroids   Other species of *Cuthona* (X.26)

XI Bryozoans

Bryozoans can be identified accurately only by using a low-power stereomicroscope. The form of the colony is useful, but the important characters are those of the single units, or zooids (literally, little animals) of the colony. The upper, or frontal, surface of each zooid has an orifice through which a bell of feeding tentacles is everted. The frontal surface may be membranous or calcified, i.e. strengthened by a calcium carbonate skeleton, and must be seen clearly in order to identify the species. This key includes all species likely to be found on intertidal algae. However, numerous other species occur on hard substrata; these may be identified using the four Linnean Society Synopses of British Bryozoa (Ryland & Hayward, 1977; Hayward & Ryland, 1979, 1985; Hayward, 1985).

7 Identification

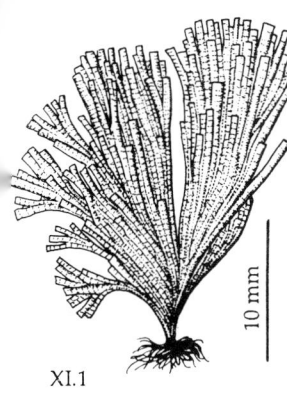

XI.1

1 Colonies encrusting; forming flat sheets, small patches or mounds, or sometimes larger, fleshy and knobbed growths from an extensive incrustation   2
− Colonies not encrusting, erect or hanging (XI.1); forming thick, bushy or spiralled growths up to 20 mm high, or diffuse and spindly, or in dense, tangled tufts, or with small (less than 1 mm), upright cylindrical zooids attached to a creeping stolon   21

2 Colonies uncalcified, smooth and gelatinous, or rather fleshy with numerous papillae on the surface, or with thick, brown spines   3
− Colonies partly or wholly calcified; sometimes with the frontal surface of the zooid membranous, to a greater or lesser extent, but with at least the vertical walls of the zooids calcified   5

3 Each zooid bearing few or many sharply pointed chitinous spines. Colony purplish brown. Predominantly on *Fucus serratus* (pl. 3.4); also *Gigartina* and *Chondrus*
                       Flustrellidra hispida (Fabricius) (XI.2)
− Zooids without spines   4

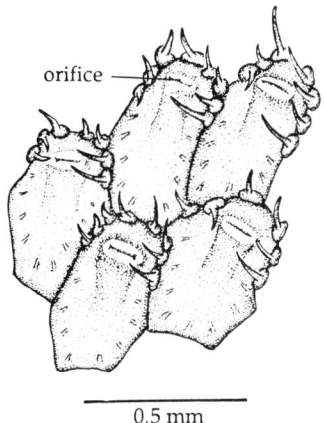

XI.2

4 Colony surface with numerous rounded or conical papillae, visible with a hand lens; velvety to touch. Forms flat incrustations, or sometimes short, finger-like erect lobes. Predominantly on *Fucus serratus*, also *Chondrus* and *Gigartina*, more rarely on other small red algae
                       Alcyonidium hirsutum (Fleming) (XI.3)
− Colony surface flat, smooth to touch. Forms a thin, gelatinous incrustation on *Fucus serratus*, very rarely on other algae
                       Alcyonidium gelatinosum (Linn.) (XI.4)

5 Colony white, disc-shaped or lobed, heavily calcified. Zooids flat or erect, tubular, sometimes sealed at the ends by a porous lid when old, but without a hinged flap (operculum). Zooids may be interspersed with open-ended, calcified polygons   6
− Not as described. Zooids always with a hinged flap, or operculum, in a calcified or membranous frontal wall   8

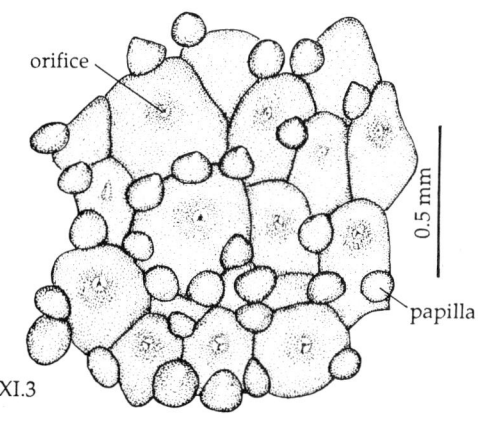

XI.3

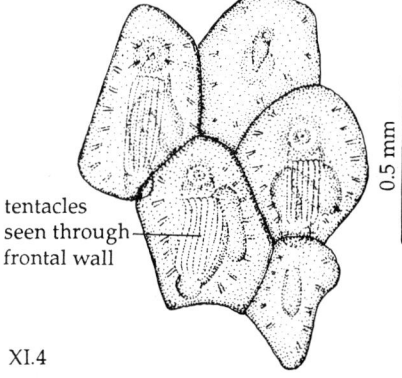

XI.4

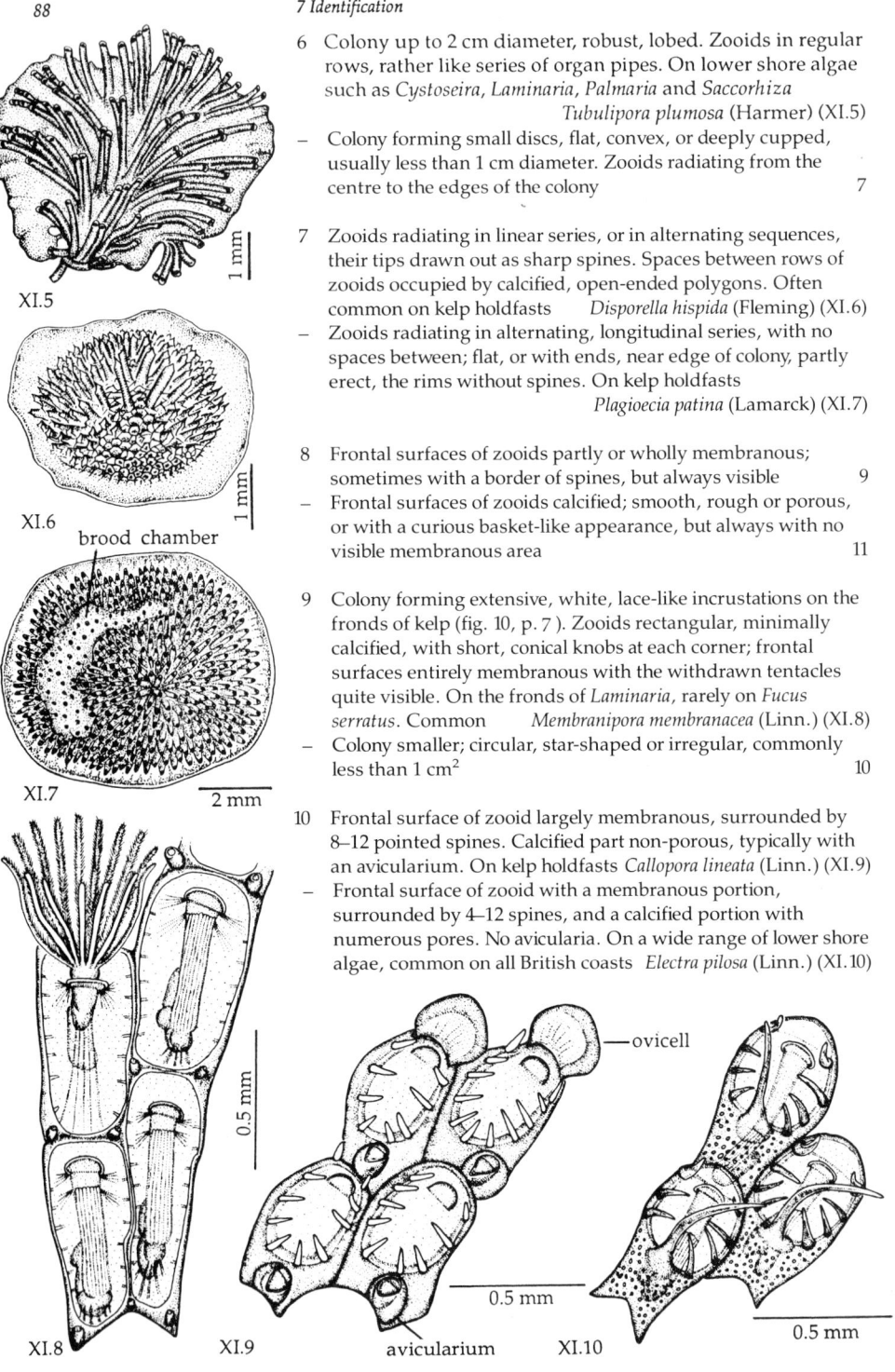

7 Identification

6   Colony up to 2 cm diameter, robust, lobed. Zooids in regular rows, rather like series of organ pipes. On lower shore algae such as *Cystoseira*, *Laminaria*, *Palmaria* and *Saccorhiza*   *Tubulipora plumosa* (Harmer) (XI.5)

—   Colony forming small discs, flat, convex, or deeply cupped, usually less than 1 cm diameter. Zooids radiating from the centre to the edges of the colony   7

7   Zooids radiating in linear series, or in alternating sequences, their tips drawn out as sharp spines. Spaces between rows of zooids occupied by calcified, open-ended polygons. Often common on kelp holdfasts   *Disporella hispida* (Fleming) (XI.6)

—   Zooids radiating in alternating, longitudinal series, with no spaces between; flat, or with ends, near edge of colony, partly erect, the rims without spines. On kelp holdfasts   *Plagioecia patina* (Lamarck) (XI.7)

8   Frontal surfaces of zooids partly or wholly membranous; sometimes with a border of spines, but always visible   9

—   Frontal surfaces of zooids calcified; smooth, rough or porous, or with a curious basket-like appearance, but always with no visible membranous area   11

9   Colony forming extensive, white, lace-like incrustations on the fronds of kelp (fig. 10, p. 7). Zooids rectangular, minimally calcified, with short, conical knobs at each corner; frontal surfaces entirely membranous with the withdrawn tentacles quite visible. On the fronds of *Laminaria*, rarely on *Fucus serratus*. Common   *Membranipora membranacea* (Linn.) (XI.8)

—   Colony smaller; circular, star-shaped or irregular, commonly less than 1 cm^2 10

10   Frontal surface of zooid largely membranous, surrounded by 8–12 pointed spines. Calcified part non-porous, typically with an avicularium. On kelp holdfasts *Callopora lineata* (Linn.) (XI.9)

—   Frontal surface of zooid with a membranous portion, surrounded by 4–12 spines, and a calcified portion with numerous pores. No avicularia. On a wide range of lower shore algae, common on all British coasts   *Electra pilosa* (Linn.) (XI.10)

7 Identification

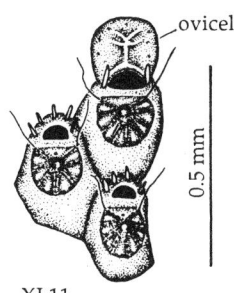

XI.11

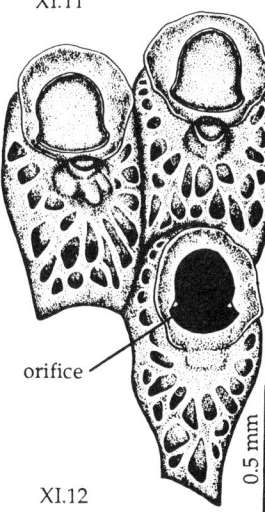

XI.12

11 Colonies small, usually less than 5 mm diameter, rounded; frontal surface of zooid consisting of fused, flattened ribs with pores between them, appearing like a small basket
   North coasts, on *Laminaria* fronds
   Cribrilina annulata (Fabricius) (pl. 8.1)
   Southwest coasts, on small red algae
   Puellina gattyae (Landsborough) (XI.11)
— Not as described   12

12 Zooids rather large, more than 0.6 mm long, heavily calcified. A single avicularium on the lower side of the orifice (XI.12)   13
— Not as described   14

13 Orifice of zooid bell-shaped, the rim typically flared. Frontal wall with large, deeply sunk pores over the entire surface. On *Laminaria* holdfasts   *Cryptosula pallasiana* (Moll) (XI.12)
— Orifice of zooid oval, without a flared rim. Frontal wall with well-marked marginal pores, but without pores in the centre. On *Laminaria* holdfasts   *Umbonula littoralis* (Hastings) (pl. 8.4)

14 Colony knobbly, developing as low convex mounds, and often forming massive nodules or cylindrical incrustations up to 2 cm long. Zooids with small avicularia, and interspersed with large spoon-shaped avicularia (XI.13, XI.14, XI.15)   15
— Not as described   16

15 Zooids with small paired avicularia flanking the orifice. Ovicell (larval brood chamber) with a crescent-shaped sieve-like frontal area. On kelp holdfasts
   Celleporina hassallii (Johnston) (XI.13)
— Each zooid with a single avicularium beside the orifice. Ovicell with large or small pores but not with a frontal crescent
   On kelp holdfasts   *Turbicellepora avicularis* (Hincks) (XI.14)
   Forming cylinders on lower shore fucoids, Scilly Isles only
   Turbicellepora magnicostata (Barroso) (XI.15)

16 Colonies lightly calcified, translucent; zooids without avicularia
                                                                      17
— Calcification thick, opaque. Zooids bearing avicularia   18

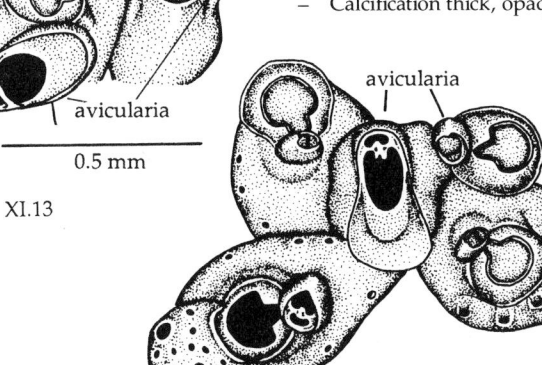

XI.13

XI.14

XI.15

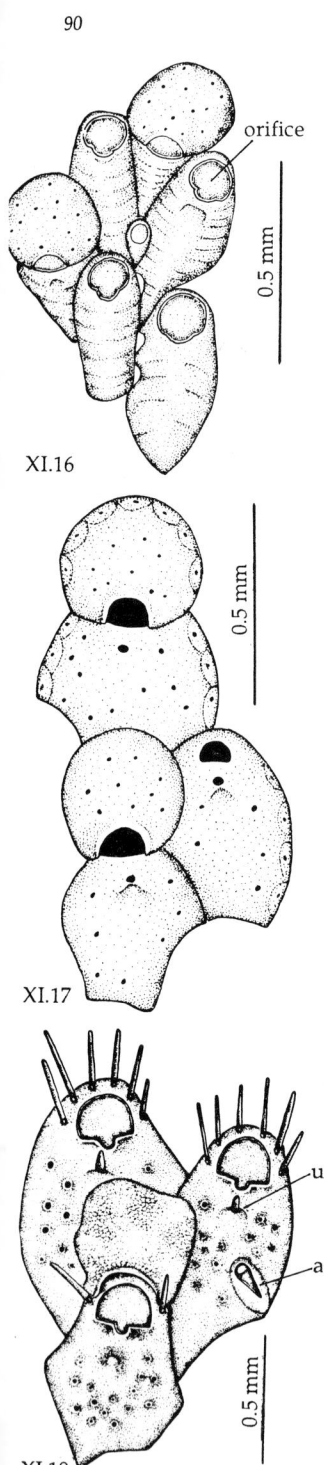

17 Zooids of several different sizes, ovicells (larval brood chambers) borne only by small, specialised female zooids. Orifice of zooid with concave lower edge. Frontal walls without pores. On *Laminaria* holdfast, stipe and frond, also on numerous small red algae. Widespread and common
    Celleporella hyalina (Linn.) (XI.16)

– Zooids of one size only, ovicells borne by ordinary zooids. Orifice of zooid with straight lower edge. Frontal wall with scattered pores. On small red algae, south and west shores only   *Haplopoma impressum* (Audouin) (XI.17)

18 Frontal wall of zooid with scattered small pores, and with a single, conspicuous crescentic pore, larger than the rest, in the middle. With one or two avicularia at the sides of each zooid. On kelp holdfasts   *Microporella ciliata* (Pallas) (pl. 8.6)

– Not as described   19

19 Frontal wall with a prominent umbo, often surmounted by a chitinous spine. Avicularia, when present, situated on the sides of the zooids, away from the orifice. On *Laminaria* holdfasts, and on other large algae *Phaeostachys spinifera* (Johnston) (XI.18)

– Without an umbo. Avicularia situated on each side of zooid orifice. Spines, when present, around the upper edges of the orifice only   20

20 Frontal wall coarse and nodular, with marginal pores only. Orifice with well-developed spines on the upper edge, the lower edge developed into a blunt tooth. On kelp holdfasts
    Escharoides coccineus (Abildgaard) (XI.19)

– Frontal wall with closely spaced pores over its entire surface. No spines. Orifice with a clearly marked notch (sinus) on lower border. On kelp holdfasts, beneath *Himanthalia* buttons, rarely on other large algae
    Schizoporella unicornis (Johnston in Wood) (pl. 8.5)

21 Colony calcified, white, delicate; comprising chains of horn-shaped zooids in single rows, or a slender, cylindrical stolon bearing individual, straight or curved, upright cylinders, each with a membranous region at the end   22

– Not as described   23

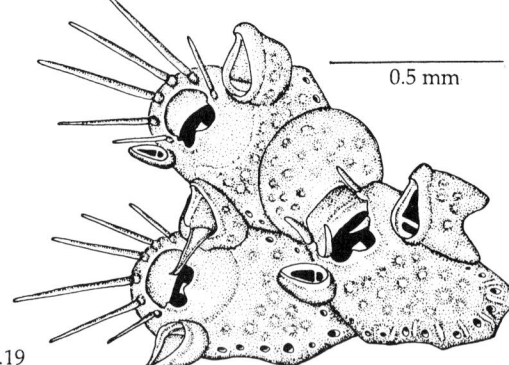

7 Identification

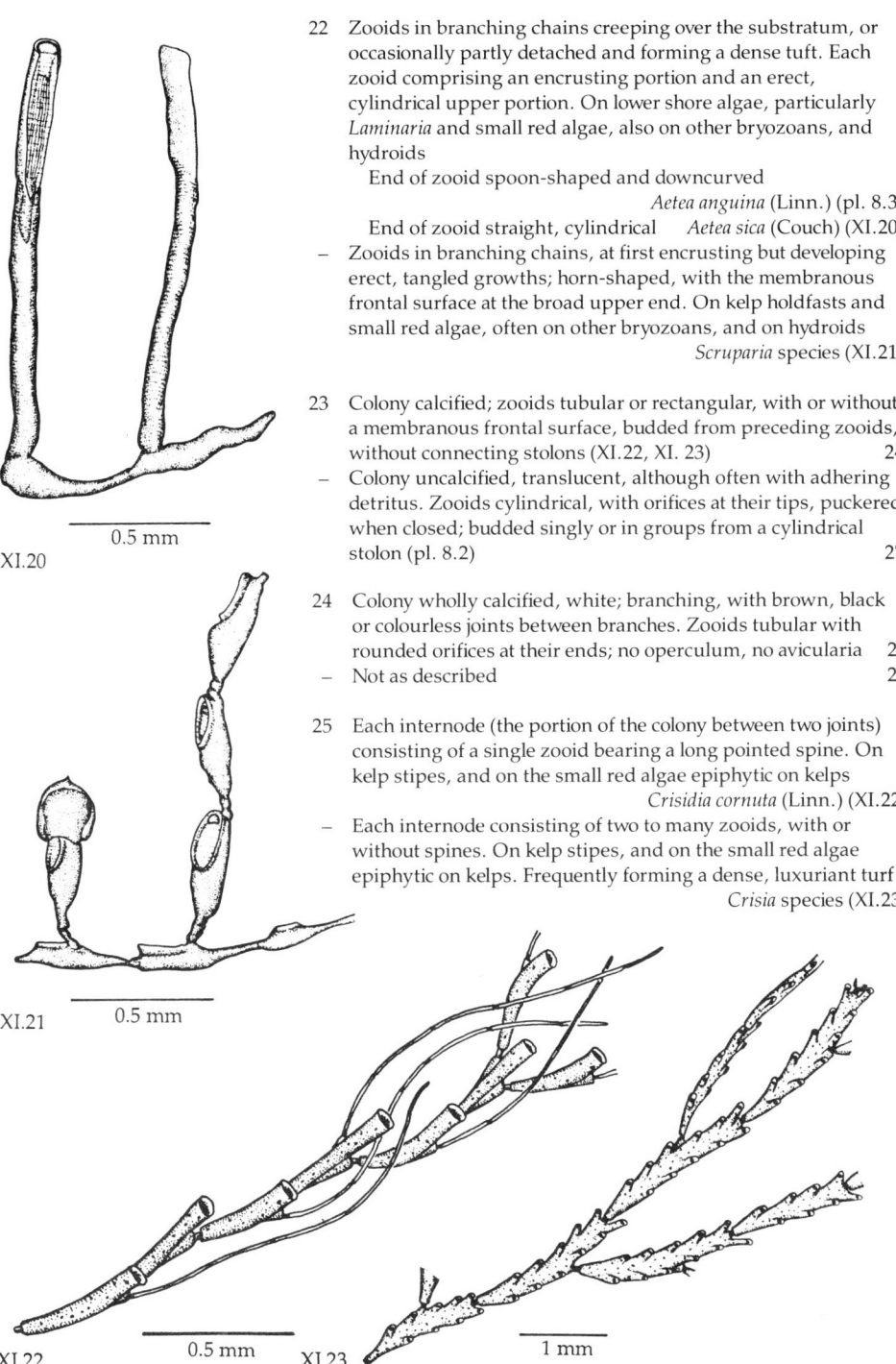

22   Zooids in branching chains creeping over the substratum, or occasionally partly detached and forming a dense tuft. Each zooid comprising an encrusting portion and an erect, cylindrical upper portion. On lower shore algae, particularly *Laminaria* and small red algae, also on other bryozoans, and hydroids
      End of zooid spoon-shaped and downcurved
                                      Aetea anguina (Linn.) (pl. 8.3)
      End of zooid straight, cylindrical   *Aetea sica* (Couch) (XI.20)
–   Zooids in branching chains, at first encrusting but developing erect, tangled growths; horn-shaped, with the membranous frontal surface at the broad upper end. On kelp holdfasts and small red algae, often on other bryozoans, and on hydroids
                                                     Scruparia species (XI.21)

23   Colony calcified; zooids tubular or rectangular, with or without a membranous frontal surface, budded from preceding zooids, without connecting stolons (XI.22, XI. 23)   24
–   Colony uncalcified, translucent, although often with adhering detritus. Zooids cylindrical, with orifices at their tips, puckered when closed; budded singly or in groups from a cylindrical stolon (pl. 8.2)   27

24   Colony wholly calcified, white; branching, with brown, black or colourless joints between branches. Zooids tubular with rounded orifices at their ends; no operculum, no avicularia   25
–   Not as described   26

25   Each internode (the portion of the colony between two joints) consisting of a single zooid bearing a long pointed spine. On kelp stipes, and on the small red algae epiphytic on kelps
                                                  Crisidia cornuta (Linn.) (XI.22)
–   Each internode consisting of two to many zooids, with or without spines. On kelp stipes, and on the small red algae epiphytic on kelps. Frequently forming a dense, luxuriant turf
                                                        Crisia species (XI.23)

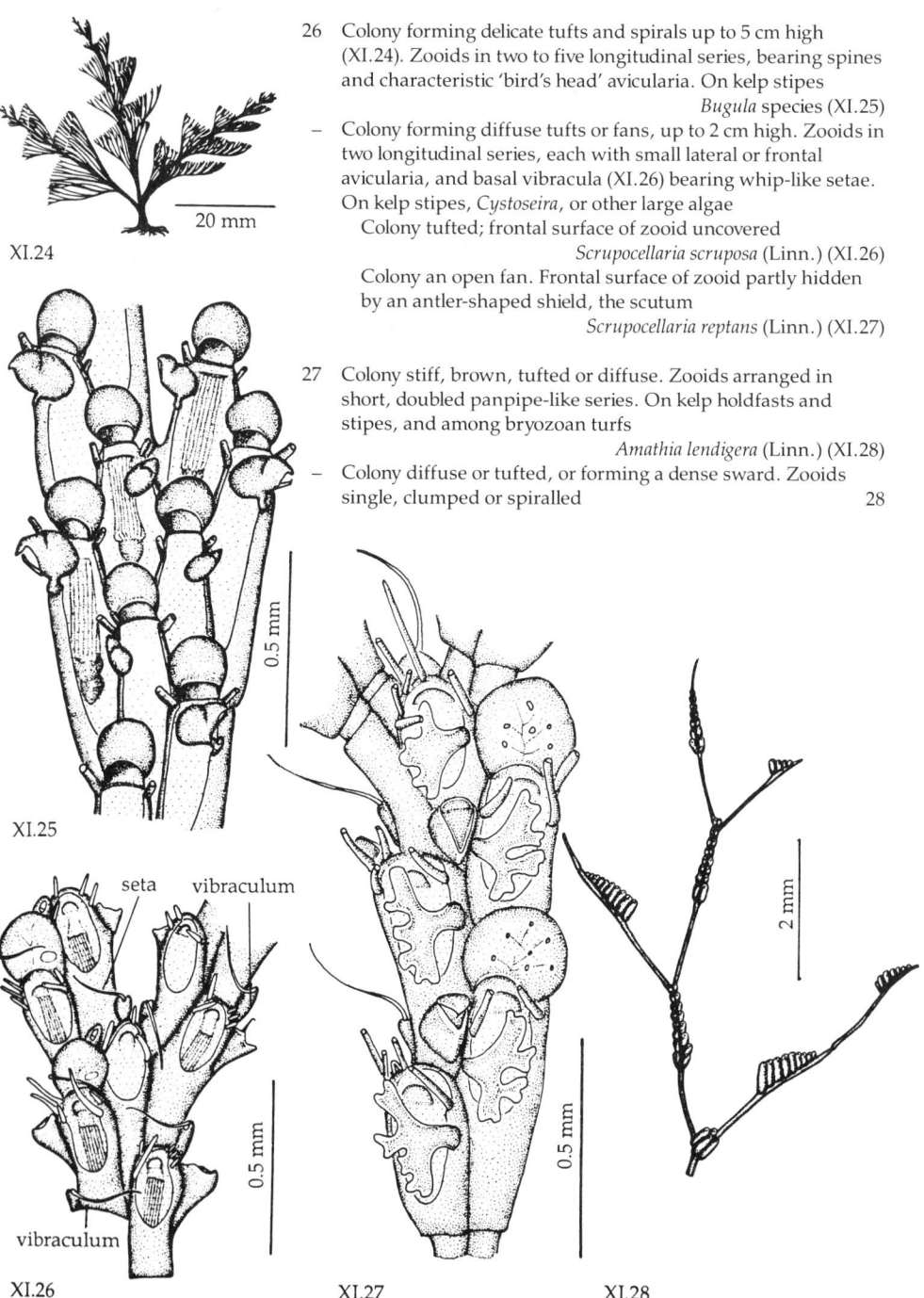

26 Colony forming delicate tufts and spirals up to 5 cm high (XI.24). Zooids in two to five longitudinal series, bearing spines and characteristic 'bird's head' avicularia. On kelp stipes
    Bugula species (XI.25)
– Colony forming diffuse tufts or fans, up to 2 cm high. Zooids in two longitudinal series, each with small lateral or frontal avicularia, and basal vibracula (XI.26) bearing whip-like setae. On kelp stipes, *Cystoseira*, or other large algae
    Colony tufted; frontal surface of zooid uncovered
    Scrupocellaria scruposa (Linn.) (XI.26)
    Colony an open fan. Frontal surface of zooid partly hidden by an antler-shaped shield, the scutum
    Scrupocellaria reptans (Linn.) (XI.27)

27 Colony stiff, brown, tufted or diffuse. Zooids arranged in short, doubled panpipe-like series. On kelp holdfasts and stipes, and among bryozoan turfs
    Amathia lendigera (Linn.) (XI.28)
– Colony diffuse or tufted, or forming a dense sward. Zooids single, clumped or spiralled   28

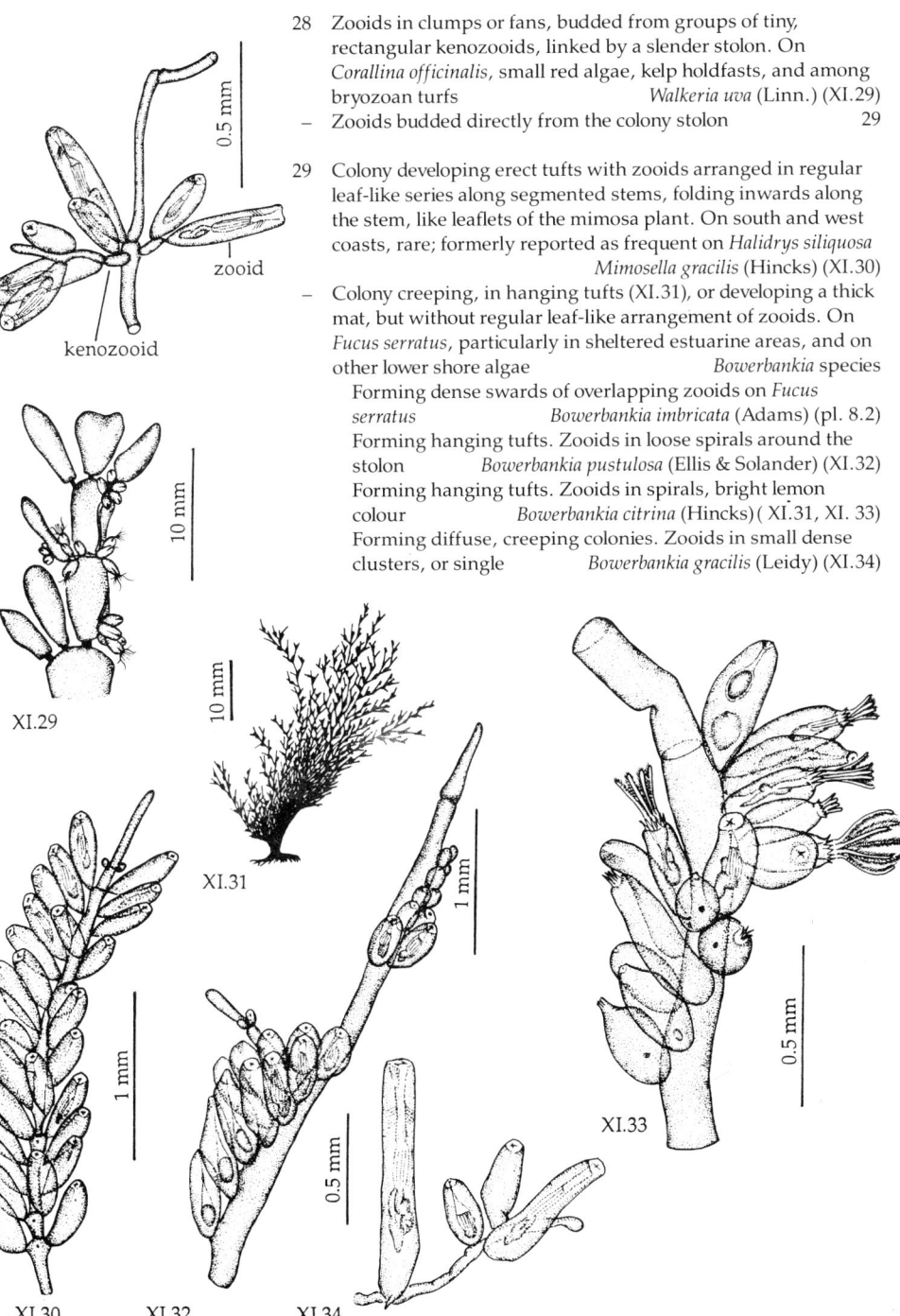

28 Zooids in clumps or fans, budded from groups of tiny, rectangular kenozooids, linked by a slender stolon. On *Corallina officinalis*, small red algae, kelp holdfasts, and among bryozoan turfs *Walkeria uva* (Linn.) (XI.29)
– Zooids budded directly from the colony stolon 29

29 Colony developing erect tufts with zooids arranged in regular leaf-like series along segmented stems, folding inwards along the stem, like leaflets of the mimosa plant. On south and west coasts, rare; formerly reported as frequent on *Halidrys siliquosa* *Mimosella gracilis* (Hincks) (XI.30)
– Colony creeping, in hanging tufts (XI.31), or developing a thick mat, but without regular leaf-like arrangement of zooids. On *Fucus serratus*, particularly in sheltered estuarine areas, and on other lower shore algae *Bowerbankia* species
Forming dense swards of overlapping zooids on *Fucus serratus* *Bowerbankia imbricata* (Adams) (pl. 8.2)
Forming hanging tufts. Zooids in loose spirals around the stolon *Bowerbankia pustulosa* (Ellis & Solander) (XI.32)
Forming hanging tufts. Zooids in spirals, bright lemon colour *Bowerbankia citrina* (Hincks) (XI.31, XI. 33)
Forming diffuse, creeping colonies. Zooids in small dense clusters, or single *Bowerbankia gracilis* (Leidy) (XI.34)

XII Ascidians

Ascidians, or sea squirts, may be quite abundant in some marine habitats. In sheltered estuaries, for example, hard substrata or substantial algal clumps may be heavily colonised by ascidians. The larger solitary species are easily identified, but many of the smaller solitary and colonial species can only be identified by dissection. A Linnean Society Synopsis (Millar, 1970) provides keys to all British species, and more detailed accounts are given by Berrill (1950).

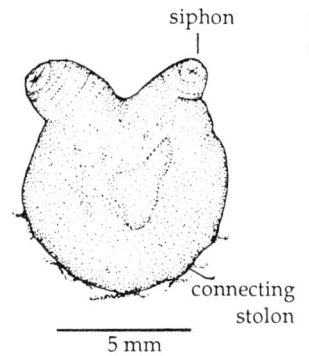

XII.1

1 Solitary ascidians, occurring singly or in dense clumps, but each individual attached independently to the substratum, or fused to its neighbours only at the base or by stolonic connection; each with two siphons at the end or one siphon at the end and one at the sides (XII.1)   2
- Colonial ascidians, forming flat sheets, with the zooids arranged in star-shaped or chain-like series; or short, stalked, clubbed or mushroom-shaped growths, with the zooids arranged in vertical series (XII.9–XII.12)   9

2 Squat, oval, hard, with paired siphons on upper surface. Dull-coloured, the outer wall rough and usually incorporating silt or sand grains. Sometimes found in dense groups but individuals detachable, with no fusion between neighbours. Up to 3 cm long. Lower shore; primarily on hard substrata, but may occur on kelp holdfasts   *Molgula* species (XII.1)
- Not as described   3

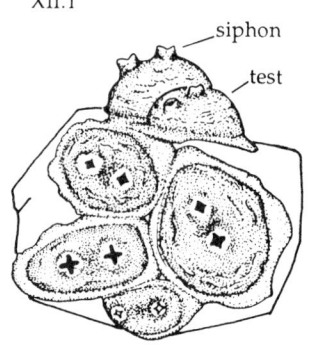

XII.2

3 Squat, oval, hard, with paired siphons on upper surface. Bright red to reddish brown, the test smooth or wrinkled, but not incorporating sand grains. May be found singly, but typically in dense, knobbly clumps, the basal portion of each individual fused with that of its neighbours. Rarely longer than 1 cm   4
- Individuals elongate, oval, often very large; sometimes in dense clumps, may be linked by slender stolons, but not fused   5

4 Apertures of siphons typically prominent, rims with four lobes (visible when expanded). Occurs singly, in small groups or dense knobbed masses; individuals up to 1 cm high. Lower shore, on hard substrata, sometimes on kelp holdfasts. All coasts, common   *Dendrodoa grossularia* (Van Beneden) (XII.2)
- Apertures of siphons with four-lobed rims; tightly withdrawn and inconspicuous out of water. In flat, dense groups; smaller than preceding species, up to 5 mm high. Lower shore, on hard substrata, frequently on *Laminaria* holdfasts and stipes. Southwest and west coasts only
  Distomus variolosus (Gaertner) (XII.3)

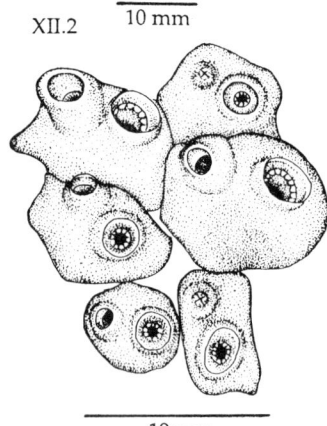

XII.3

7 Identification

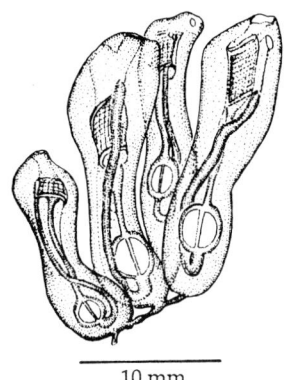

XII.4

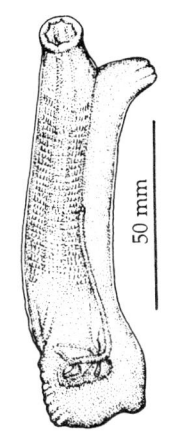

XII.5

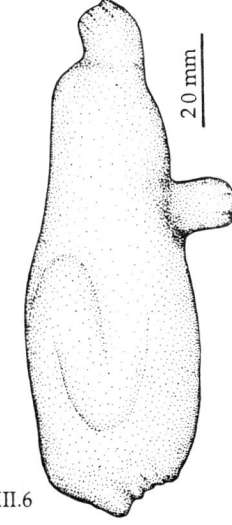

XII.6

5   Delicate, gelatinous, stalked zooids in small clumps, linked by an inconspicuous basal stolon. Up to 2 cm high; transparent with bright yellow, pink, brown or white internal pigmentation. Lower shore, on kelp holdfasts and larger fucoid algae. Widespread and common   *Clavelina lepadiformis* (Muller) (XII.4)
–   Individuals often densely clumped, but without basal stolons. Opaque or translucent; greyish, or tinted red or green, but without vivid internal pigmentation   6

6   Soft, gelatinous, readily contractile; cylindrical, attached at base of cylinder with closely spaced siphons at top. Transparent, greyish, pale green or yellowish, with bright red spots around apertures of siphons. Up to 10 cm long. Lower shore and subtidal, often common in sheltered estuaries, docks and harbours   *Ciona intestinalis* (Linn.) (XII.5)
–   Body stiff, or even hard, siphons more widely spaced   7

7   Body attached to substratum at base, rough-textured; opaque, colourless, greyish or pink tinted. Elongate, cylindrical; inhalant siphon at the top, exhalant siphon at the side, situated about one-third body length from tip. Up to 5 cm high, sometimes larger. South and west coasts, often abundant in estuaries; lower shore, on hard substrata, fucoids and kelps
Ascidiella aspersa (Muller) (XII.6)
–   Body attached to substratum by its side   8

8   Body stiff, rough-textured close to siphons, elsewhere smooth or wrinkled. Inhalant siphon at the top, exhalant siphon near the top, both rather elongate. Up to 5 cm long. Transparent, greyish or with pink tints. Lower shore and shallow sublittoral, on hard substrata and kelp holdfasts. All coasts
Ascidiella scabra (Muller) (XII.7)
–   Body firm, rough-surfaced, often encrusted with other organisms. Inhalant siphon at the top, exhalant siphon about halfway along body length. Up to 20 cm long. Only partly translucent, may be greyish or greenish, but typically red or reddish brown. Lower shore and sublittoral, south and west coasts. On hard substrata, often on kelp holdfasts
Ascidia mentula (Muller) (XII.8)

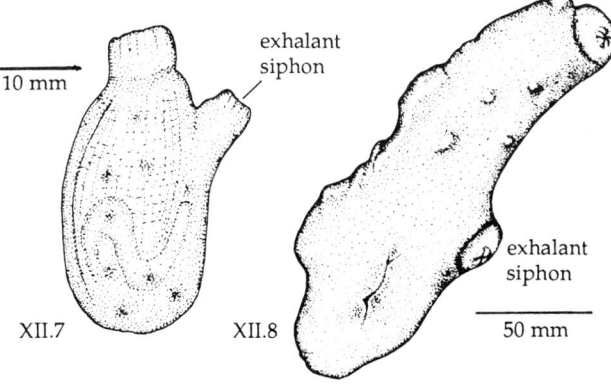

XII.7   XII.8

7 Identification

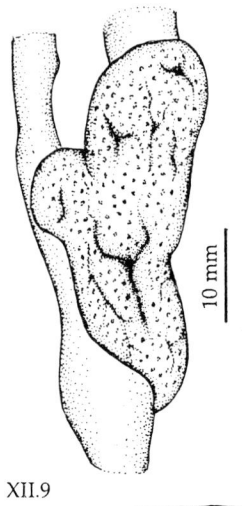

XII.9

XII.10

9 Colony forming broad, gelatinous or coarse-textured sheets, often brightly coloured 10
– Colony forming stalked, cylindrical, oval or mushroom-shaped clumps 12

10 Colony developing a flat, coarse-surfaced, bluish grey or light purple incrustation on kelp stipes. Individual zooids not readily recognisable *Didemnum candidum* (Savigny) (XII.9)
– Colony developing smooth, gelatinous sheets, often brightly coloured and variegated. On hard substrata, but frequent also on *Laminaria* fronds and large lower shore fucoids 11

11 Zooids arranged in star-shaped groups around common exhalant siphons. Richly coloured: orange, yellow, green, blue or violet, with creamy yellow or red spots
Botryllus schlosseri (Pallas) (pl. 6.3)
– Zooids in long branching chains, sometimes rather disordered. Predominantly red, orange or yellow
Botrylloides leachi (Savigny) (pl. 6.1)

12 Colonies in the form of short, flat-topped mushrooms, widely spaced and linked by slender stolons. Up to 1 cm high. Transparent, yellow or orange, often with red spots. South and west coasts. Lower shore, frequently on kelp holdfasts
Sidnyum turbinatum (Savigny) (XII.10)
– Not as described 13

13 Forming short, globular, flat-topped growths, closely bunched together. Common exhalant siphons visible. Yellow or brownish yellow, up to 3 cm high. South and west coasts. Lower shore, on hard substrata and kelp holdfasts
Polyclinum aurantium (Milne-Edwards) (XII.11)
– Elongate, clubbed growths forming loose clumps. Yellow or orange, often with red spots. Up to 4 cm long. Lower shore, on hard substrata and kelp holdfasts *Aplidium* species (XII.12)

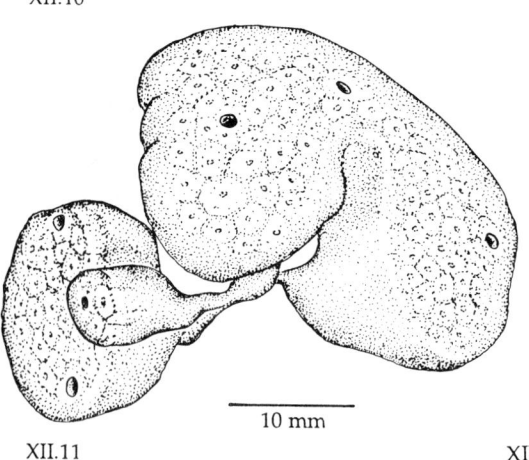

XII.11

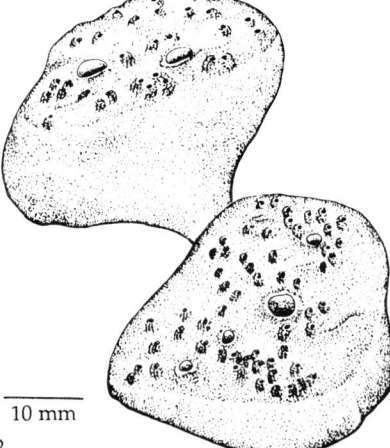

XII.12

8 Techniques

The study of intertidal seaweed faunas requires no special techniques beyond care and attention to detail in the collection of data. The approach is obviously determined by the objective of the study. Comparison of epiphyte communities at different tidal levels or between different shores is best made within the framework of a formal survey in which tidal levels can be established for each algal population, shore profiles constructed and the degree of exposure assessed. Techniques for shore surveying and sampling are described in detail in Price and others (1980).

More detailed, and perhaps more interesting, work may be conducted with the epiphyte populations of single algal species. The value of this kind of investigation depends very much on sampling procedures, which should be carefully thought out beforehand. Whole plants should be collected by bagging, carefully drawing a polythene bag over the alga and closing it around the holdfast before severing its attachment. Silt loads may be estimated (see p. 25) by washing each plant in a known volume of seawater and then filtering subsamples of the washings, and drying and weighing the filter papers. The more robust brown seaweeds may be frozen for later measurement of plant characteristics, such as weight, length, frond width, etc., but the animal communities should be dealt with soon after collection. Some species can be preserved in 70% alcohol for later examination but others may yield more useful information while still alive. Molluscs may be narcotised prior to preservation by immersion overnight in a 7% solution of magnesium chloride (see Smaldon & Lee (1979) for methods of narcotisation); bryozoans will relax if left overnight in seawater to which menthol crystals have been added.

Reference specimens of soft-bodied animals may be stored permanently in 5% seawater formalin; shelled animals and sponges are better stored in 70% alcohol. Data labels should give details of locality, shore level or depth, habitat and date. Permanent labels may be made using goatskin parchment, or plastic-covered paper, and indian ink, and will not deteriorate in either alcohol or formalin.

The measurement of turbulence using gypsum balls (p. 25) is a simple technique developed originally by Muus (1968). Plaster of Paris may be cast in any useful sphere - split squash balls would seem to be ideal - and a piece of stout string inserted before the plaster hardens. The balls are dried to constant weight and then tied to selected plants.

Turbulence is estimated as percentage weight loss over a tidal cycle and may be compared between different sites, between plants of different bushiness, or between different levels on the same plant.

Long-term monitoring of single algal populations and their epiphyte communities is one of the most urgent needs in this field of marine ecology. Individual plants, particularly of the fucoid algae, are easily marked using loops of thin plastic-covered wire loosely tied around the stipe. Different colour combinations allow large numbers of plants to be individually recognised. Data related to plant growth rate, reproduction and defoliation, and growth and recruitment of epiphyte communities may be measured directly in the field, or recorded photographically. Transplantation of plants, or rather of heavy rocks with well-attached plants, is a potentially interesting way of studying the effects of environmental change on plant and epiphyte characteristics.

It is conventional wisdom that sampling and experimentation should be designed with analysis in mind. There are numerous books, at all levels, dealing with the statistical treatment of biological data and each investigator will have an opinion on that most suitable to the project in hand. However, elegant statistical treatment will not compensate for poor sampling procedures, and unnecessarily detailed analysis may only serve to cloud perfectly obvious conclusions. In principle, it is perhaps best to make use of those methods which help to simplify particularly complex problems. For example, the Shannon-Wiener diversity index used by Edwards (1980) is one way of comparing holdfast faunas. Worked examples of this index are given by Wratten & Fry (1980). The problems of measuring ecological diversity are discussed in detail by Pielou (1975, 1977).

Seaweed faunas are ideal subjects for biological and ecological research projects. Their often dense populations are ready sources of abundant material, although wasteful sampling should be avoided, and they are usually sufficiently constant from year to year to permit long-term monitoring. Many of the sessile species, particularly among the bryozoans and spirorbid tubeworms, tend to release larvae epidemically over fairly short periods of time. Species of *Alcyonidium* release larvae in February-March (or September-October in some populations), while *Flustrellidra* releases larvae in June-July. Breeding colonies may be recognised by dense creamy-white patches of advanced embryos, visible to the unaided eye. Samples kept damp, cool and in the dark for about 12 hours after collection will freely release larvae on immersion in clean seawater at daylight illumination, and the larvae may be used, preferably within a few hours of release, for settlement experiments. Spirorbid tubeworms are brooders and the worms must be removed from their tubes before larvae can be collected. Apart from epiphytic species, and their predators, numerous marine invertebrates, and some fish, deposit egg masses or capsules on seaweeds. In fact, the number of species is so large that a key for the identification of egg masses is not a practical proposition. However, careful observation over periods of time will allow correlation between increased population levels of certain species and the appearance of particular types of egg mass or capsule. In many cases it will be possible to rear these in marine aquaria and record early development stages. Plankton-feeding larvae are difficult, in some cases impossible, to rear to the juvenile stage. However, Todd (1981) provides detailed information on techniques for rearing sea slug larvae, many of which have prolonged planktonic stages. Yolk-feeding larvae, many of which have long incubation periods in substantial, easily handled egg cases, often have only a short free-swimming stage, if they have one at all, and these should be good candidates for experimental rearing.

Publication of results is sometimes possible through local natural history societies, some of which produce periodical journals, and contact with an officer of the society will yield information on these. *Field Studies*, the journal of the Field Studies Council, also publishes original articles on regional survey work, as well as the excellent AIDGAP keys. There are a number of professional journals, including the *Journal of the Marine Biological Association of the United Kingdom*, the *Journal of Experimental Marine Biology and Ecology* and *Marine Biology*, and many more specialised ones. Submission of manuscripts to professional journals should only be considered after discussion with a professional biologist. The Marine Biological Association of the United Kingdom, at Plymouth, and the Scottish Marine Biological Association, at Oban, are the premier British professional societies and an introduction to a member of either of them may result in permission to consult their extensive libraries.

The Field Studies Council maintains a number of first-class field centres ideally situated for marine ecological studies, such as Dale Fort, Pembrokeshire, and Slapton Ley, Devon. The serious student is strongly recommended also to enquire about local groups. Many Naturalists' Trusts and local diving clubs have active research or monitoring programmes and usually welcome new enthusiasts. Nationally, the Nature Conservancy Council and the

Marine Conservation Society also initiate and support regional research and monitoring groups. Finally, the burgeoning interest in marine biology over the past two decades has revealed many threatened habitats. There are now numerous protected coastal reserves: before you commence work on a particular beach *always* check with the local Naturalists' Trust that you are not about to disturb a conserved area.

Some useful addresses

Marine Biological Association of the United Kingdom,
The Laboratory, Citadel Hill, Plymouth PL1 2PB.

Scottish Marine Biological Association,
Dunstaffnage Marine Research Laboratory, PO Box 3, Oban, Argyll PA34 4AD.

Information about field courses is available from:
The Information Officer, Field Studies Council, Preston Montford, Montford Bridge, Shrewsbury SY4 1HW.

Addresses of local naturalists' trusts are available from:
Royal Society for Nature Conservation, 22 The Green, Nettleham, Lincoln LN2 2NR.

Linnean Society Synopses and AIDGAP keys are available from:
The Richmond Publishing Co. Ltd., Orchard Road, Richmond, Surrey TW9 4PD.

References and further reading

Finding books

Some of the books and journals listed here will be unavailable in local and school libraries. It is possible to make arrangements to see or borrow such works by seeking permission to visit the library of a local university, or by asking your local public library to borrow the work (or a photocopy of it) for you via the British Library, Document Supply Centre. This may take several weeks, and it is important to present your librarian with a reference that is correct in every detail. References are acceptable in the form given here, namely the author's name, the date of publication, followed by (for a book) the title and publisher or (for a journal article) the title of the article, the journal title, the volume number, and the first and last pages of the article.

References

Al-Ogily, S. & Knight-Jones, E.W. (1977). Anti-fouling role of antibiotics produced by marine algae and bryozoans. *Nature (London)*, **265**, 728-9.

André, M. (1946). Halacariens marins. *Faune de France*, **46**. (In French.)

Ballantine, W.J. (1961). A biologically-defined exposure scale for the comparative description of rocky shores. *Field Studies*, **1**(3), 1-19.

Barnes, R.S.K. & Hughes, R.N. (1982). *An Introduction to Marine Ecology*. Oxford: Blackwell Scientific Publications.

Barrett, J.H. & Yonge, C.M. (1958). *Pocket Guide to the Seashore*. London: Collins.

Bergquist, P.R. (1978). *Sponges*. London: Hutchinson.

Berrill, N.J. (1950). *The Tunicata with an Account of the British Species*. London: Ray Society.

Bryce, D. & Hobart, A. (1972). The biology and identification of the larvae of the Chironomidae (Diptera) *Entomologist's Gazette*, **23**, 175-217.

Chadwick, S.R. & Thorpe, J.P. (1981). An investigation of some aspects of bryozoan predation by dorid nudibranchs (Mollusca: Opisthobranchia). In *Recent and Fossil Bryozoa*, ed. G.P. Larwood & C. Nielsen, pp. 51-8. Fredensborg: Olsen & Olsen.

Chevreux, E. & Fage, L. (1925). Amphipodes. *Faune de France*, **9**. (In French.)

Colman, J. (1940). On the faunas inhabiting intertidal seaweeds. *Journal of the Marine Biological Association of the United Kingdom*, **24**, 129-83.

Conover, J.T. & Sieburth, J. McN. (1966). Effect of tannins excreted from Phaeophyta on planktonic animal survival in tide pools. *Proceedings of the International Seaweed Symposium*, **5**, 99-100.

Cornelius, P.F.S. (1975). A revision of the species of Lafoeidae and Haleciidae (Coelenterata: Hydroida) recorded from Britain and nearby seas. *Bulletin of the British Museum (Natural History), Zoology*, **28**(8), 376-426.

Cornelius, P.F.S. (1979). A revision of the species of Sertulariidae (Coelenterata: Hydroida) recorded from Britain and nearby seas. *Bulletin of the British Museum (Natural History), Zoology*, **34**(6), 243-321.

Cornelius, P.F.S. (1982). Hydroids and medusae of the family Campanulariidae recorded from the eastern North Atlantic, with a world synopsis of genera. *Bulletin of the British Museum (Natural History), Zoology*, **42**(2), 37-139.

Cranston, P.S. (1982). A key to the larvae of the British Orthocladiinae (Chironomidae).

Freshwater Biological Association, Scientific Publications, **45**, 1-152.

Daly, J.M. (1978). The annual cycle and the short term periodicity of breeding in a Northumberland population of *Spirorbis spirorbis* (Polychaeta: Serpulidae). *Journal of the Marine Biological Association of the United Kingdom*, **58**, 161-76.

Dickinson, C.I. (1963). *British Seaweeds*. Kew Series. Eyre and Spottiswoode, London.

Doyle, R.W. (1975). Settlement of planktonic larvae: a theory of habitat selection in varying environments. *American Naturalist*, **109**, 113-26.

Ebling, F.J., Kitching, J.A., Purchon, R.A. & Bassindale, R. (1948). The ecology of the Lough Ine Rapids with special reference to water currents. II. The fauna of the *Saccorhiza* canopy. *Journal of Animal Ecology*, **17**, 223-44.

Edwards, A. (1980). Ecological studies of the kelp, *Laminaria hyperborea*, and its associated fauna in south-west Ireland. *Ophelia*, **19**(1), 47-60.

Fauvel, P. (1923). *Faune de France*, vol. 5, *Polychètes errantes*. Paris: Fédération Française des Sociétés de Science Naturelle. (In French.)

Fauvel, P. (1927). *Faune de France*, vol. 16, *Polychètes sédentaires*. Paris: Fédération Francaise des Sociétés de Science Naturelle. (In French.)

Fretter, V. & Graham, A. (1962). *British Prosobranch Mollusca. Their Functional Anatomy and Ecology*. Ray Society Monographs no. 144. London: Ray Society.

Fretter, V. & Manly, R. (1977). Algal associations of *Tricolia pullus, Lacuna vincta* and *Cerithiopsis tubercularis* (Gastropoda) with special reference to the settlement of their larvae. *Journal of the Marine Biological Association of the United Kingdom*, **57**, 999-1017.

Gallop, A., Bartrop, J. & Smith, D.C. (1980). The biology of chloroplast acquisition by *Elysia viridis*. *Proceedings of the Royal Society of London, Series B*, **207**, 335-49.

George, J.D. & Hartmann-Schroder, G. (1985). *British Polychaetes: Amphinomida, Spintherida and Eunicida*. Synopses of the British Fauna (n.s.) no. 32. Leiden: E.J. Brill for the Linnean Society of London and the Estuarine and Brackish-water Sciences Association.

Gibson, R. (1982). *British Nemerteans*. Synopses of the British Fauna (n.s.) no. 24. Cambridge University Press for the Linnean Society of London.

Graham, A. (1971). *British Prosobranchs*. Synopses of the British Fauna (n.s.) no. 2. Academic Press for the Linnean Society of London.

Grahame, J. (1977). Reproductive effort and *r*- and *K*-selection in two species of *Lacuna* (Gastropoda: Prosobranchia). *Marine Biology*, **40**(3), 217-24.

Green, J. & Macquitty, M.(1987). *Halacarid Mites*. Synopses of the British Fauna (n.s.) no. 36. E. J. Brill for the Linnean Society of London and The Estuarine and Brackish-water Sciences Association. .

Harrison, R.J. (1944). *Caprellidea*. Linnean Society Synopses of the British Fauna no. 2. London: Linnean Society.

Hayward, P.J. (1973). Preliminary observations on settlement and growth in populations of *Alcyonidium hirsutum* (Fleming). In *Living and Fossil Bryozoa*, ed. G.P. Larwood, pp. 107-13. London: Academic Press.

Hayward, P.J. (1980). Invertebrate epiphytes of coastal marine algae. In *The Shore Environment*, vol. 2, *Ecosystems*, ed. J.H.Price, D.E.G. Irvine & W.F. Farnham, pp. 761-87. London: Academic Press.

Hayward, P.J. (1985). *Ctenostome Bryozoans*. Synopses of the British Fauna (n.s.) no. 34. Leiden: E.J. Brill for the Linnean Society of London and the Estuarine and Brackish-water Sciences Association.

Hayward, P.J. & Harvey, P.H. (1974). The distribution of settled larvae of the bryozoans *Alcyonidium hirsutum* (Fleming) and *Alcyonidium polyoum* (Hassall) on

Fucus serratus L. *Journal of the Marine Biological Association of the United Kingdom*, **54**, 665-76.

Hayward, P.J. & Ryland, J.S. (1979). *British Ascophoran Bryozoans*. Synopses of the British Fauna (n.s.) no. 14. London: Academic Press for the Linnean Society of London and the Estuarine and Brackish-water Sciences Association.

Hayward, P.J. & Ryland, J.S. (1985). *Cyclostome Bryozoans*. Synopses of the British Fauna (n.s.) no. 35. Leiden: E.J. Brill for the Linnean Society of London and the Estuarine and Brackish-water Sciences Association.

Hiscock, S. (1979). A field key to the British brown seaweeds (Phaeophyta). *Field Studies*, **5**, 1-44.

Hiscock, S. (1986). A field key to the British red seaweeds (Rhodophyta). *AIDGAP Field Studies Council Occasional Publication, no.13*.

Hornsey, I.S. & Hide, D. (1976). The production of antimicrobial compounds by British marine algae. III. Distribution of antimicrobial activity within the algal thallus. *British Phycological Journal*, **11**, 175-81.

Jensen, K. (1975). Food preference and food consumption in relation to growth of *Limapontia capitata* (Opisthobranchia, Sacoglossa). *Ophelia*, **14**, 1-14.

Jones, D.J. (1971). Ecological studies on macroinvertebrate populations associated with polluted kelp forests in the North Sea. *Helgolander Wissenschaftliche Meeresuntersuchungen*, **23**, 248-60.

Jones, D.J. (1973). Variation in the trophic structure and species composition of some invertebrate communities in polluted kelp forests in the North Sea. *Marine Biology*, **20**, 351-65.

Kain, J.M. (1963) Aspects of the biology of *Laminaria hyperborea*. II. Age, weight and length. *Journal of the Marine Biological Association of the United Kingdom*, **43**, 129-51.

Kain, J.M.(1971). Synopsis of the biological data on *Laminaria hyperborea*. FAO *Fisheries Synopsis*, **87**, 1-74.

Kain, J.M. & Svensden, P. (1969). A note of the behaviour of *Patina pellucida* in Britain and Norway. *Sarsia*, **38**, 25-30.

King, P.E. (1974). *British Sea Spiders*. Synopses òf the British Fauna (n.s.) no. 5. London: Academic Press for the Linnean Society of London.

King, P.E. (1986). Sea spiders. A revised key to the adults of littoral Pycnogonida in the British Isles. *Field Studies* , **6**, 493-516 (An AIDGAP key).

Knight, M. & Parke, M. (1950). A biological study of *Fucus vesiculosus* L. and F. *serratus* L. *Journal of the Marine Biological Association of the United Kingdom*, **29**, 439-514.

Knight-Jones, E.W., Bailey, J.H. & Isaac, M.J. (1971). Choice of algae by larvae of *Spirorbis*, particularly of *Spirorbis spirorbis*. In *Fourth European Marine Biology Symposium*, ed. D.J. Crisp, pp. 89-104. Cambridge University Press.

Knight-Jones, P. & Knight-Jones, E.W. (1977). Taxonomy and ecology of British Spirorbidae (Polychaeta). *Journal of the Marine Biological Association of the United Kingdom*, **57**, 453-500.

Lewis, J.R. (1964). *The Ecology of Rocky Shores*. London: English Universities Press.

Lewis, J.R. (1980). Objectives in littoral ecology - a personal viewpoint. In *The Shore Environment*, vol.1, *Methods*, ed. J.H. Price, D.E.G. Irvine & W.F. Farnham, pp. 1-18. London: Academic Press.

Lincoln, R.J. (1979). *British Marine Amphipoda: Gammaridea*. London: British Museum (Natural History).

Meadows, P.S. & Campbell, J.I. (1972). Habitat selection by aquatic invertebrates.

Advances in Marine Biology, **10**, 271-382.
Meadows, P.S. & Williams, G.B. (1963). Settlement of *Spirorbis borealis* (Daudin) larvae on surfaces bearing films of micro-organisms. *Nature (London)*, **198**, 610-11.
Millar, R.H. (1970). *British Ascidians*. Synopses of the British Fauna (n.s.) no. 1. London: Academic Press for the Linnean Society of London.
Moore, P.G. (1974). The kelp fauna of northeast Britain. III. Qualitative and quantitative ordinations, and the utility of a multivariate approach. *Journal of Experimental Marine Biology and Ecology*, **16**, 257-300.
Moore, P.G. (1978). Turbidity and kelp holdfast Amphipoda.I. Wales and S.W. England. *Journal of Experimental Marine Biology and Ecology*, **32**, 53-96.
Mortensen, T. (1977). *Handbook of the Echinoderms of the British Isles*. Rotterdam: Backhuys. (Reprint of 1927 edition by Oxford University Press.)
Muus, B.J. (1968). A field method for measuring 'exposure' by means of plaster balls. *Sarsia*, **34**, 61-8.
Myers, A.A. & Southgate, T. (1980). Artificial substrates as a means of monitoring rocky shore cryptofauna. *Journal of the Marine Biological Association of the United Kingdom*, **60**, 963-75.
Naylor, E. (1972). *British Marine Isopods*. Synopses of the British Fauna (n.s.) no. 3. London: Academic Press for the Linnean Society of London.
Pielou, E.C. (1975). *Ecological Diversity*. New York: Wiley.
Pielou, E.C. (1977). *Mathematical Ecology*. New York: Wiley.
Platt, H.M. & Warwick, R.M. (1983). *Free-living Marine Nematodes*, Part 1, *British Enoplids*. Synopses of the British Fauna (n.s.) no. 28. Cambridge University Press for the Linnean Society of London and the Estuarine and Brackish-water Sciences Association.
Price, J.H., Irvine, D.E.G. & Farnham, W.F. (1980). *The Shore Environment*: vol. 1, pp. 1-321; vol. 2, pp. 322-945. London: Academic Press.
Ryland, J.S. (1959). Experiments on the selection of algal substrates by polyzoan larvae. *Journal of Experimental Biology*, **36**, 613-31.
Ryland, J.S. (1962). The association between Polyzoa and algal substrata. *Journal of Animal Ecology*, **31**, 331-8.
Ryland, J.S. & Hayward, P.J. (1977). *British Anascan Bryozoans*. Synopses of the British Fauna (n.s.) no. 10. London: Academic Press for the Linnean Society of London.
Ryland, J.S. & Nelson-Smith, A. (1975). Littoral and benthic investigations on the west coast of Ireland. IV. Some shores in counties Clare and Galway. *Proceedings of the Royal Irish Academy, Series B*, **75**, 245-66.
Sara, M. (1974). *Catalogue of Main Marine Fouling Organisms*, vol. 5, *Marine Sponges*. Paris: OECD.
Seed, R. & O'Connor, R.J. (1981). Community organisation in marine algal epifaunas. *Annual Review of Ecology and Systematics*, **12**, 49-74.
Smaldon, G. & Lee, E.W. (1979). A synopsis of methods for the narcotization of marine invertebrates. *Royal Scottish Museum Information Series, Natural History*, **6**, 96 pp.
Smith, D.A.S. (1973). The population biology of *Lacuna pallidula* (Da Costa) and *Lacuna vincta* (Montagu) in north east England. *Journal of the Marine Biological Association of the United Kingdom*, **53**, 493-520.
Southgate, T. (1982). The biology of *Barleeia unifasciata* (Gastropoda: Prosobranchia) in red algal turfs in S.W. Ireland. *Journal of the Marine Biological Association of the United Kingdom*, **62**, 461-8.
Southward, A.J. (1976). On the taxonomic status and distibution of *Chthamalus*

stellatus (Cirripedia) in the north east Atlantic region: with a key to the common intertidal barnacles of Britain. *Journal of the Marine Biological Association of the United Kingdom*, 56, 1007-28.

Stebbing, A.R.D. (1972). Preferential settlement of a bryozoan and serpulid larvae on the younger parts of *Laminaria* fronds. *Journal of the Marine Biological Association of the United Kingdom*, 52, 765-72.

Stebbing, A.R.D. (1973). Competition for space between the epiphytes of *Fucus serratus* L. *Journal of the Marine Biological Association of the United Kingdom*, 53, 247-61.

Tebble, N. (1976). *British Bivalve Seashells: A Handbook for Identification*, 2nd edition. Edinburgh: HMSO for the Royal Scottish Museum.

Tebble, N. & Chambers, S. (1982). *Polychaetes from Scottish Waters*, part 1, *Family Polynoidae*. Edinburgh: Royal Scottish Museum.

Thomas, M.L.H. & Page, F.H. (1983). Grazing by the gastropod, *Lacuna vincta*, in the lower intertidal area at Musquash Head, New Brunswick, Canada. *Journal of the Marine Biological Association of the United Kingdom*, 63, 725-36.

Thompson, T.E. (1958). The natural history, embryology, larval biology and post-larval development of *Adalaria proxima* (Alder and Hancock). *Philosophical Transactions of the Royal Society of London, Series B*, 242, pp. 1-58.

Thompson, T.E. (1966). Studies on the reproduction of *Archidoris pseudoargus* (Rapp). *Philosophical Transactions of the Royal Society of London, Series B*, 250, 343-75.

Thompson, T.E. (1976). *Biology of Opisthobranch Molluscs*, vol. I. Ray Society Monographs no. 151. London: Ray Society.

Thompson, T.E. & Brown, G.H. (1976). *British Opisthobranch Molluscs*. Synopses of the British Fauna (n.s.) no. 8. London: Academic Press for the Linnean Society of London.

Thompson, T.E. & Brown, G.H. (1984). *Biology of Opisthobranch Molluscs*, vol. II. Ray Society Monographs no. 156. London: Ray Society.

Todd, C.D. (1978). Changes in spatial pattern of an intertidal population of the nudibranch mollusc *Onchidoris muricata* in relation to life-cycle, mortality, and environmental heterogeneity. *Journal of Animal Ecology*, 47, 189-203.

Todd, C.D. (1981). The ecology of nudibranch molluscs. *Oceanography and Marine Biology Annual Review*, 19, 141-234.

Vahl, O. (1971). Growth and density of *Patina pellucida* (L.) (Gastropoda: Prosobranchia) on *Laminaria hyperborea*. *Ophelia*, 9, 31-50.

Wigham, G.D. (1975). The biology and ecology of *Rissoa parva* (da Costa). (Gastropoda: Prosobranchia.) *Journal of the Marine Biological Association of the United Kingdom*, 55, 45-67.

Williams, G.B. (1964). The effect of extracts of *Fucus serratus* in promoting the settlement of larvae of *Spirorbis borealis* (Polychaeta). *Journal of the Marine Biological Association of the United Kingdom*, 44, 397-414.

Wratten, S.D. & Fry, G.L.A. (1980). *Field and Laboratory Exercises in Ecology*. London: Edward Arnold.

Index

Acanthodoris pilosa 33, 84
Acari 43
Achelia 32
   A. echinata 73
   A. hispida 73
   A. longipes 32, 73
Acidostoma sarsi 63
Adalaria proxima 33, 84
   feeding 34
   life cycle 34
Aetea anguina 91
Aetea sica 91
Aglaeophenia pluma 50
Alcyonidium 83, 84
   A. gelatinosum 6, 33, 34, 36, 37, 87
     larval settlement 22
   A. hirsutum 6, 19, 33, 73, 87
     competition 25
     larval settlement 22, 25
     life cycle 25
     predation of 35
Alderia modesta 82
Alentia gelatinosa 55
Alvania punctura 79
Amathia lendigera 92
Amphilochus manudens 63
Amphipods 30, 41, 45, 62-71
Amphisbetia operculata 52
Amphitrite 56
Ampithoe gammaroides 67
Ampithoe rubricata 67
Anomia ephippium 74
Anoplodactylus 32, 73
Anthura gracilis 60
Apherusa bispinosa 65
Apherusa jurinei 65
Aplidium 96
Archidoris pseudoargus 33, 35, 83
Ascidia mentula 95
Ascidians 44, 47, 94-96
Ascidiella aspersa 95
Ascidiella scabra 36, 95
Ascophyllum nodosum 2, 4, 6, 7, 19, 49, 61
Atylus swammerdami 65

Balanus perforatus 40, 41
Barentsia gracilis 40
Barleeia unifasciata 9, 80
   life cycle 17
Barnacles 40, 46
Berthella plumula 81
Bivalves 43, 46, 74-76

Bladder wrack 6
Blue-rayed limpet 8, 9
Botrylloides leachi 33, 77, 83,96
Botryllus schlosseri 33, 36, 77, 83, 96
Bowerbankia 71, 73
   B. citrina 93
   B. gracilis 93
   B. imbricata 7, 32, 93
   B. pustulosa 93
Branchiomma bombyx 28, 56
Brittle stars 44, 47
Brown algae 2
Bryopsis 82
Bryozoa 44, 47, 86-93
Bugula 92

Callithamnion 15, 16
Callopora lineata 88
Calma glaucoides 85
Calycella syringa 51
Campecopea hirsuta 60
Caprella acanthifera 70
Caprella fretensis 71
Caprella linearis 71
Caprella septentrionalis 71
Caprellids 42, 45, 70-71
Carcinus maenas 28
Castalia punctata 58
Cellaria fistulosa 36
Cellaria sinuosa 36
Celleporella hyalina 6, 20, 36, 90
Celleporina hassallii 89
Ceramium sesquipedale 15, 16
Cerithiopsis tubercularis 14, 15, 77
Chaetogammarus marinus 65
Cheirocratus sundevallii 66
Chironomids 43
Chitons 46
Chlorophyceae 2
Chondrus crispus 2, 7, 9, 15, 19, 54, 73, 78, 80, 87
Cingulopsis fulgida 80
Ciona intestinalis 95
Circeis armoricana 54
Cladophora 2, 16, 18, 61, 82
   C. rupestris 15
Clava multicornis 49
Clavelina lepadiformis 95
Clunio 43
Clytia hemisphaerica 51
Codium 18, 81, 82
Copepods 40, 45
Corallina officinalis 9, 15, 16, 17, 54, 77, 93
Corophium acutum 66

Corophium bonellii 30, 31, 67
Corophium sextonae 67
Coryne 49
Coryphella 33, 85
Cribrilina annulata 89
Crisia 91
Crisidia cornuta 91
Cryptosula pallasiana 84, 89
Cucumaria 28, 44
Cuthona nana 86
Cystoseira tamariscifolia 4, 54, 88, 92
Cythere albomaculata 42

Delesseria 54, 82
Dendrodoa grossularia 33, 94
Dendronotus frondosus 33, 85
Detritivores 11-18
Dexamine spinosa 70
Diatoms 10, 15
Didemnum candidum 96
Didemnum maculosum 24
Disporella hispida 88
Distomus variolosus 94
Dogwhelk 32
Doto coronata 33, 85
Doto dunnei 33
Doto millbayana 33
Dynamena pumila 7, 19, 32, 33, 52, 71, 73
Dynamene bidentata 28, 60

Echinoderms 44, 47
Elasmopus rapax 66
Electra pilosa 6, 7, 24, 33, 34, 36, 37, 84
Elysia viridis 82
   habitat selection, 18
   nutrition 18
Endeis 73
Enteromorpha 16, 82
Entoprocts 40, 47
Epiphyte, definition of 3
Erato voluta 77
Erichthonius 67
Escharoides coccineus 84, 90
Eubranchus 33, 86
Eulalia sanguinea 59
Eulalia viridis 59
Eusyllis 58
Exposure 3, 7, 9
   measurement 97

Facelina coronata 85
Filter feeders 19-27
Flustrellidra hispida 6, 19, 24, 25, 33, 34, 36, 73, 84, 87
Fucus 2, 16, 61, 81

Fucus edentatus,
  grazing effects on 14
Fucus serratus 4, 6, 7, 9, 15,
  48, 52, 53, 87, 88, 93
  epifauna 19, 20, 24, 25, 32,
  33, 34, 35
  grazing effects on 11-14
  larval settlement on 21-23
  life cycle 23
Fucus spiralis 4, 5, 6
Fucus vesiculosus 4, 6, 11, 53
  antibacterial activity 27
  larval settlement on 21
Furcellaria 54

Gammarella fucicola 66
Gammarellus angulosus 65
Gammarellus homari 65
Gammaropsis maculata 69
Gelidium sesquipedale 9, 15
Gibbula 11
Gibbula cineraria 78
Gibbula umbilicalis 11, 78
Gigartina stellata 2, 4, 7, 9, 15,
  16, 19, 54, 87
Gitana sarsi 63
Gnathia maxillaris 60
Goniodoris castanea 33, 83
Goniodoris nodosa 33, 83
  diet of 36, 37
Gonothyraea loveni 33,52
Grantia compressa 24
Green algae 2
Griffithsia 82
Grubea 58

Habitat selection by larvae
  20-24, 26, 27
Halacaridae 33, 43
Halacarellus basteri 43
Halecium halecinum 33
Halichondria panicea 14, 33,
  35, 48, 83
Halidrys siliquosa 11, 50, 61, 93
Haplopoma impressum 90
Harmothoe 28
  H. extenuata 55
  H. imbricata 55
  H. impar 55
Harpacticus uniremis 40
Hartlaubella gelatinosa 51
Herbivores 11-18
Hermaea bifida 82
Hermaea dendritica 82
Heteranomia squamula 31, 74
Hiatella arctica 74, 76
Himanthalia elongata 26, 54
Hyale nilssoni 69

Hyale pontica 69
Hydrallmania falcata 33
Hydroides norvegica 53
Hydroids 39, 47, 49-52
Hymeniacidon perleve 14, 33, 48

Idotea 6
  I. baltica 61, 62
  I. chelipes 62
  I. emarginata 61, 61
  I. granulosa 7, 61
  I. neglecta 62
  I. pelagica 61
Iphimedia minuta 64
Iphimedia obesa 64
Ischyrocerus anguipes 68
Isopods 41, 45, 60-62

Jaera 62
Janira maculosa 62
Janua pagenstecheri 54
Jassa falcata 68
Jorunna tomentosa 33, 83

Kefersteinia cirrata 58
Kellia suborbicularis 28, 75
Kelp 7
  holdfast fauna 8, 20, 28-31
  holdfast volume,
    estimating 29
Kirchenpaueria pinnata 33, 50
Knotted wrack 7

Lacuna pallidula 7, 11, 14, 80
  life cycle 12, 13
Lacuna parva 80
Lacuna vincta 7, 11, 14, 15,
  16, 80
  life cycle 12, 13
Lamellibranchs 43, 74-76
Laminaria 2, 11, 50, 51, 54,
  55, 57, 60, 77, 88, 89, 90
  91, 94, 96
  antibacterial activity 27
  holdfast fauna 7, 8, 29-31
  holdfast volume,
    estimating 29
Laminaria digitata 4, 19, 20, 50
  larval settlement on 22, 28
Laminaria hyperborea 19
  ageing 9, 29
  holdfast volume,
    estimating 29
  Patina infestation 8
Laminaria saccharina 20, 28, 54
Laomedea flexuosa 52
Larvae, experiments with
  98, 99

Lasaea rubra 76
Laurencia pinnatifida 15, 16,
  17, 19
Lembos websteri 30, 68
Lepadogaster 85
Lepidochitona cinereus 44
Lepidonotus clava 54
Lepidonotus squamatus 54
Leucosolenia 48
Leucothoe incisa 64
Leucothoe spinicarpa 64
Lichina pygmaea 60, 76
Liljeborgia pallida 65
Lima hians 74
Limapontia capitata 18, 82
Limapontia depressa 82
Limapontia senestra 82
Lineus longissimus 39
Littorina arcana 81
Littorina littorea 1, 11
Littorina obtusata 6, 11, 81
Littorina saxatilis 5, 81
Lomentaria articulata 9, 15,
  16, 17, 18

Megalomma vesiculosum 56
Membranipora membranacea 7,
  19, 24, 33, 34, 36, 84, 88
Microdeutopus gryllotalpa 69
Microdeutopus versiculatus 69
Microjassa cumbrensis 68
Microporella ciliata 84, 90
Microprotopus maculatus 69
Mimosella gracilis 93
Mites 33, 43, 45
Modiolus barbatus 75
Modiolus modiolus 75
Molgula 94
Monia patelliformis 31
Munna kroyeri 62
Musculus 28
  M. costulatus 75
  M. discors 75
  M. marmoratus 75
Mytilus edulis 28, 74
Myxilla incrustans 33

Nematodes 39, 46
Nemerteans 39, 46
Neodexiospira pseudocorrugata 54
Neries 28
Nereis pelagica 57
Nicolea 56
Notirus irus 76
Nucella lapillus 32
Nymphon brevirostre 71
Nymphon gracile 32, 71

Obelia	85, 86	Red algae	2	Sertularia	85
Obelia dichotoma	52	infauna	9, 14, 15, 17, 32	Sidnyum turbinatum	96
Obelia geniculata	33, 51	sediment traps	14, 15	Silt, quantification of	25
Odontosyllis	58	Rhodophyceae	2	Skeleton shrimps	42, 45, 70-71
Omalogyra atomus	78	Ribbon worms	39, 46		
Onchidoris bilamellata	84	Rissoa	9	Skeneopsis planorbis	78
Onchidoris muricata	33, 84	Rissoa guerini	80	Specimens, care of	97
diet	36, 37	Rissoa lilacina	79	Sphaerosyllis	58
population study	35	Rissoa membranacea	80	Sphenia binghami	76
Onchidoris pusilla	84	Rissoa parva	79	Spiralled wrack	5
Onoba semicostata	79	adhesive gland	16	Spirorbis corallinae	54
Ophiothrix fragilis	28, 44	life cycle	16	Spirorbis inornatus	54
Opisthobranchs	43, 44, 47, 81-86	population structure	18	habitat selection by	26
		Rissoella	80	Spirorbis rupestris	53
Orchomene humilis	63	Round worms	39, 46	Spirorbis spirorbis	6, 7, 19, 24
Orchomene nana	63	Rostanga rubra	83	larval settlement	20, 21, 22
Orthopyxis integra	51	Runcina coronata	81	Sponges	39, 47, 48
Ostracods	42, 46			Staphyllococcus, suppression	
		Sabellaria	28	by seaweed tissue	26
Palio dubia	84	Saccorhiza	54, 85, 88	Stenothoe marina	64
Palmaria palmata	11, 88	Sandhoppers	41, 45	Stenothoe monoculoides	64
Parajassa pelagica	68	Sargassum muticum	27	Sugar Kelp	20
Patella	11, 77	Sarsia eximia	49	Sunamphitoe pelagica	69
Patella vulgata	2	Schizoporella	84	Sycon ciliatum	24
Patina pellucida	8, 9, 28, 77	Schizoporella unicornis	90	Syllis	57
var. laevis	8, 77	Scruparia	91	Synisoma acuminatum	61
Peacock weed	4	Scrupocellaria, larval settlement			
Pedicellina cernua	40		22	Tergipes tergipes	86
Pelvetia canaliculata	1, 2, 4, 5, 6, 81	Scrupocellaria reptans	36, 92	Tonicella rubra	44
		Scrupocellaria scruposa	92	Tricolia pullus	9, 14, 16, 78
Perinereis cultrifera	57	Scypha ciliata	19, 48	life cycle	15
Phaeophyceae	2	Scypha compressa	19, 48	Tritaeta gibbosa	70
Phaeostachys spinifera	90	Sea cucumbers	44, 47	Trivia arctica	77
Phtisica marina	70	Sea slugs	18, 43, 44, 47, 81-86	Trivia monacha	77
Phyllodoce lamelligera	59	population study	35	Tubularia	85, 86
Phymatolithon	53	preferred diets of	33, 34	Tubulipora plumosa	88
Pilumnus hirtellus	28	prey selection	36	Turbicellepora avicularis	89
Pisidia longicornis	28	Sea snails	43, 46, 77-81	Turbicellepora magnicostata	89
Plagioecia patina	88	Sea spiders	10, 32, 42, 45, 71-73	Turbidity, effects on holdfast	
Platynereis dumerilii	57			fauna	30, 31
Plumaria elegans	15, 16	Sea squirts	44, 47, 94-96	Turbulence, measurement of	25
Plumularia setacea	33, 50	Seaweeds,		Turtonia minuta	76
Podocerus variegatus	66	antibacterial activity of			
Pollution, effects on holdfast			26, 27	Ulva	78
fauna	30, 31	effects of grazers on	11-14	Umbonula littoralis	84, 89
Polycera quadrilineata	33, 84	habitat selection by			
diet	36, 37	epifauna	5, 6	Vaucheria	82
Polychaetes	40, 46, 53-59	population of animals on		Veligers, habitat selection by	
Polyclinum aurantium	24, 96		4, 5		16
Polysiphonia lanosa	4, 61	resources for animals	1, 2, 3	Verruca stroemia	28, 40, 41
Pomatoceros triqueter	31, 53	technique for marking	98		
Predators	32-37	zonation of	1-4	Walkeria uva	93
Prosobranchs	43, 46, 77-81	Serrated wrack	6		
Pseudoprotella phasma	70	Serpula vermicularis	53		
Puellina gattyae	89	Sertularella polyzonias	52		
Pycnogonids	42, 45, 71-73	Sertularella rugosa	52		